An Introduction to Building Procurement Systems, 2nd edition

Jack W. E. Masterman

London and New York

First published 2002 by Spon Press
11 New Fetter Lane, London EC4P 4EE

Simultaneously published in the USA and Canada
by Spon Press
29 West 35th Street, New York, NY 10001

Spon Press is an imprint of the Taylor & Francis Group

© 2002 Jack W E Masterman

Typeset in Sabon by
Prepress Projects Ltd, Perth, Scotland
Printed and bound in Great Britain by
St Edmundsbury Press, Bury St Edmunds, Suffolk

British Library Cataloguing in Publication Data
A catalogue record for this book is available from the British Library

Library of Congress Cataloging in Publication Data
A catalog record has been requested

ISBN 0–415–24642 –3 (pbk)
ISBN 0–415–24641–5 (hbk)

An Introduction to Building
Procurement Systems, 2nd edition

To my brilliant grandchildren
Victoria and Michael
with much love

Contents

Figures

Tables

Acknowledgements

Grateful acknowledgement is made to the following publishers and authors for permission to quote or reproduce extracts from the works listed below.

- Bennett, J. and Grice, A. (1990) 'Procurement systems for building', in Brandon, P.S. (ed.) *Quantity Surveying Techniques: New Directions*, Oxford: Blackwell Science.
- Brandon, P.S., Basden, A., Hamilton, W.I. and Stockley, J.E. (1988) *Expert Systems: the Strategic Planning of Construction Projects*, London: Royal Institution of Chartered Surveyors.
- Building Economic Development Committee, NEDO (1967) *Action on the Banwell Report*, London: HMSO. Crown copyright is reproduced with the permission of the Controller of Her Majesty's Stationery Office.
- Building Economic Development Committee, NEDO (1983) *Faster Building for Industry,* London: HMSO. Crown copyright is reproduced with the permission of the Controller of Her Majesty's Stationery Office.
- Cole, G.A. (1990) *Management: Theory and Practice*, London: Continuum International Publishing Group.
- Construction Round Table (1995) *Thinking About Building*, London: Business Round Table.
- Flanagan, R. (1981) 'Change the system', *Building* 20 March, 28–29.
- Franks, J. (1990) *Building Procurement Systems*, Ascot: Chartered Institute of Building.
- Hewitt, R.A. (1985) 'The procurement of buildings: proposals to improve the performance of the industry', unpublished project report submitted to the College of Estate Management for the RICS Diploma in Project Management.
- Hillebrandt, P.M. (1985) *Economic Theory of the Construction Industry*, 2nd edn, Basingstoke: Macmillan.
- HM Treasury Central Unit on Purchasing (1992) *Guidance Document No. 36, Contract Strategy Selection for Major Projects*, London: HMSO. Crown copyright is reproduced with the permission of the Controller of Her Majesty's Stationery Office.
- Jones, G.P. (1984) *A New Approach to the JCT Design and Build Contract*, Harlow: Longman.

- Mohsini, R. and Davidson, C.H. (1989) 'Building procurement: key to improved performance', paper presented to the CIB International Workshop on Contractual Procedures, University of Liverpool.
- Skitmore, R.M. and Marsden, D.E. (1988) 'Which procurement system? Towards a universal procurement selection technique', *Construction Management and Economics* 6, 71–89; London: E & FN Spon.
- University of Salford (1988) *ELSIE Expert System: Users' Manual*, Salford: Royal Institution of Chartered Surveyors (the ELSIE system is now being sold, maintained and further developed by the AremisSoft Corporation, Alton, Hampshire).

My thanks are also extended to Michael P.E. Hunt for his assistance in preparing a number of the figures and tables in this work.

No doubt the industry's performance is vital to economic growth

Preface

One of the significant changes occurring in the construction industry in recent years relates to the way in which the industry now understands that the needs of clients are of paramount importance, particularly with respect to the implementation process and its effects on the client's organisation.

Despite this, the performance of construction projects, in terms of the usual criteria of time, cost and functionality/quality, continues to be less than perfect, and many clients still appear to perceive the implementation of a building project as an expensive and risky adventure with work taking a long time to begin, with delays, disputes, extra costs and the confrontational nature of the construction industry prolonging the agony.

Over the years, there has been an understandable reluctance on the part of some clients to become directly involved in the way a project is implemented and in the management of the process itself, whereas other clients, particularly those that are inexperienced in construction matters, want to become involved in the most inappropriate way. Fortunately, the current trend appears to be that more and more clients are becoming constructively concerned with the way their projects are managed.

Notwithstanding their perception of their role in the construction process, all clients will eventually face the dilemma of how best they can procure a project to minimise delay in commencement and completion, minimise risk, reduce disruption to their main business activity, obtain value for money, etc., i.e. which procurement system should they choose?

Although it is accepted that on most projects there is likely to be more than one procurement route to success and that the determination of an inappropriate procurement strategy is not the only reason for poor project performance, it is suggested that it is a significant contributory factor in the industry's apparent inability to achieve a higher level of project success and overall performance.

However, although the construction industry has been castigated by many major clients in the past – particularly over the past three decades – for poor management, there is now a large body of literature which describes the history of contemporary major military, industrial and commercial construction projects and which amply demonstrates that clients themselves

are as likely to be responsible for ineffective management of their projec
as those they employ to implement the work practically.

In particular, it has been suggested that the way in which many clients
and their advisors select their procurement strategy can be haphazard, ill-
timed and lacking in logic and discipline.

In an attempt to discover whether this is true and how the client's dilemma
is resolved in practice, I have examined and discussed most aspects of this
phenomenon with a view to drawing conclusions which, it is hoped, will be
helpful to all of the participants in that fascinating, fulfilling – but often
frustrating – series of events known as the construction process. A number
of changes have been made in both the content and structure of the second
edition, with a reorientation of some original chapters into what is considered
to be a more logical sequence and the addition of chapters on discretionary
systems and on successful procurement system selection; in addition, the
chapter on the selection of procurement systems has been enlarged to include
the results of research that has been carried out since the publication of the
first edition. I hope that these changes will assist the reader to understand
more readily the nature and attributes of building procurement systems and
the important role their proper selection plays in achieving project success.

Jack Masterman
Chester
September 2001

Introduction

This work initially addresses, in general terms, that part of the project-implementation process that begins immediately the client decides to build a new facility and ends when he/she has chosen what is perceived to be the most appropriate method of managing the design and construction of the project.

This phase of the overall process consists of carrying out a detailed assessment of the client's characteristics, the client's overall needs and objectives, identifying the specific primary and secondary objectives of the project, the risks inherent in the proposal, determining the environment in which it will be implemented and, finally, selecting the most appropriate method of procuring the project. This creates a framework, the so-called project strategy, which will facilitate and enable the remainder of the project to proceed to a successful conclusion.

It is the last element of the project strategy, the selection of the procurement system, which is examined in detail here. This examination is based upon research carried out by the author to determine whether clients procure their projects by default, by adopting a disciplined and objective course within the framework of the project strategy, by approaching the decision in a reactionary, conservative and subjective way or by using other means which might well be a combination of all, or part, of any of these three possibilities.

It is hoped that this work will provide guidance to clients and consultants regarding the most commonly available procurement systems and the principles of good procurement system selection, provide contractors with an insight into how clients should, and do, make their procurement choices and provide students with a useful guide to the vital initial stages of project implementation.

Specifically, Chapter 1 looks at the project culture, project implementation and the project strategy. Chapter 2 examines the clients of the construction industry in detail and establishes client categories, objectives and needs and looks briefly at project constraints and risks. In Chapter 3, the concept and evolution of procurement systems are discussed. Chapters 4–7 look at separated, integrated, management-orientated and discretionary procurement systems in some depth. Common variants of the different systems are examined in Chapter 8. Chapter 9 discusses theoretical and actual decision-making and the selection of procurement systems in theory and practice. Chapter 10 provides guidance on successful procurement system selection, and, finally, Chapter 11 attempts to forecast future trends in project procurement.

procurement substantially impact project culture

1 The project-implementation process

1.1 Introduction

Before examining the procurement process in detail, the latest philosophies adopted by major clients to implement their projects need to be considered because they are likely to affect substantially, at least in the immediate future, the project culture and the way in which clients manage their construction projects.

Traditional methods of implementing projects have been replaced over the past two to three decades by less conventional approaches, often incorporating more co-operative means of project implementation.

While much remains to be done to reach the level of co-operation between the demand and supply sides of the construction industry seen in other major industries, such progress as has been made is beginning to alter the management and culture of many major construction projects and is helping to build trust and increase co-operation between all members of a project team.

1.2 The project culture

Although the importance of central government as a major construction industry client has declined over the past decade, demand from this particular source remains substantial. Its construction procurement policies and concerted efforts to improve team relationships have considerable direct influence on local government as well as indirectly on the private sector and, of course, on its own considerable and varied activities.

The advent of the private finance initiative (PFI), market testing, contracting out and internal markets has meant that government departments are carrying out more of their business through procurement of goods and services rather than by direct provision. This has resulted in the devising of a new strategy to ensure that government achieves world class standards in its procurement activities [1].

In essence, the strategy consists of the following key elements:

* establishing best practice in procurement;

- achieving best value for money;
- defining departmental objectives and requirements and careful assessing of business cases, risks and contracts;
- emphasising integrated procurement processes, covering the whole cycle of acquisition and use, to ensure quality and economy over time and not just the lowest price;
- combining co-operation and competition into relationships with suppliers in order to promote continuous improvement and benefit sharing;
- the client and the industry working together to maximise the chances of project success and to enable government to help the industry to be more efficient and competitive.

Subsequent reports, such as the Cabinet Office Efficiency Unit's paper *Construction Procurement by Government* [2], have supported this new approach, having established that the construction industry was not matching best practice in other sectors of industry.

The approach adopted by the new strategy might be considered to be a restatement of guidelines which have been well defined in the past and will come as no surprise to those who have had experience on either the demand or supply side of the industry or to those who have read the Latham report [3] or earlier reports.

What is of interest, however, is the stated intention on the part of the major public client to change its relationship with the supply side of the industry by working with co-operative, competent and non-adversarial consultants and contractors in such a way as to improve the performance of all the participants in the process.

The government's philosophy therefore now appears to be that, while still wishing to ensure that value for money is achieved, there will be a positive move towards the establishment of a more co-operative and non-confrontational environment in which to carry out, by use of best practice, project implementation.

This new policy may well have stemmed, at least in part, from the less overtly stated emphasis on government continuing to act increasingly as a facilitator of capital projects through the PFI, public/private-sector partnerships and other means, thus reducing the amount of direct implementation and funding of such projects.

In terms of the public sector, the reduction of capital expenditure on major building projects and the use of contractors and others to finance, build and operate such developments means that the level of co-operation between the client and the provider and operator must be improved. This heightened level of collaboration is necessary to ensure that the facility performs as required and that it is a commercial success for all those involved. It might, of course, be that this new approach is prompted as much by the need to make a virtue out of a necessity as for any improvement in the performance of the construction industry or any particular project.

Whatever the reason for the pursuit of this new policy by the major public-sector client, a large and rapidly increasing number of influential private-sector clients are also now carrying out their projects within a similar co-operative framework in which full and sustained collaboration with all the members of the project team is a major feature of the approach.

Assuming that the demand side of the industry continues increasingly to adopt such an approach, and that the supply side responds positively, this change of attitude could ensure a gradual change from a philosophy of confrontation to one of co-operation on many major projects, with a consequent improvement in project performance.

However, such a state may only be achieved where clients have a consistent and long-term demand for construction services and where there is a coincidence of interest among the participants within the supply chain. It has been suggested [4] that only one-quarter of construction projects are likely to fall into this category and that the remainder will still need to be managed by traditional commercial methods.

Despite this stricture, it is possible that – in time – a co-operative approach, using partnering, strategic alliances and serial contracting for example, may well persuade both sides of the industry to move away from the single-project approach to building and fostering relationships to a philosophy based upon long-term programmes of construction and the adoption of wider perspectives [5].

1.3 Project implementation

Much has been written concerning construction project management and project implementation, and it is not the intention here to add to the large amount of literature that illuminates the general principles of this subject but rather to concentrate on that part of the implementation process that deals with the management of the choice of the most suitable method of procuring the project. Although the project culture and the spirit in which the project participants approach their individual tasks will affect the outcome of the implementation of the process, the overall management of the project will still need to be carried out in such a way as to ensure that the client's objectives are met by the most effective means.

Morris [6] proposes a model for the management of projects which takes into account past experience of managing major schemes in different industries, many of which are considerably more sophisticated than construction, and suggests that adopting a wider approach to the management of projects than that usually found in traditional literature and teaching will increase the chances of success.

This wider approach, while incorporating the classic methods and techniques for managing and controlling projects, stresses the need to take into account – particularly when defining the project objectives, strategy, standards, technology and design – the effect that the external environment has on these elements. There is also the need to ensure that the attitudes of all of the parties are positive and supportive and that they are motivated, treated as team members and have experienced and strong leaders.

The need for all the parties to have positive and supportive attitudes, to be well motivated and led and to work together as a team echoes the latest client philosophies, and the arguments for the adoption of such an approach in order to maximise the chances of success are irrefutable.

The effect of the external environment on the activities of organisations has been well established in management literature, as has the need for this phenomenon to be managed, but this does not detract from the importance of its application to the specific area of project management. Matters such as local community, politics, the economic environment, the timing and nature of the funding of the project, its location and physical characteristics are identified in Morris's model as being of particular importance to the successful implementation of projects.

The basic task of defining the project remains fundamental to its successful outcome, and it is contended that this can only be achieved in the case of building projects by the formulation of a project strategy which will set out the client's intentions with regard to the management of all of the activities necessary for the efficient implementation of the project.

1.4 The project strategy

The project-implementation process is usually complex, expensive and lengthy and is often carried out in a hostile environment by means of a temporary multiorganisation (TMO), i.e. the project team, the members of which are frequently unknown to each other.

Walker [7] maintains that it is the relationship between team members and their subsequent performance that is the most significant factor in determining project success *not* the prescribing of a specific method of procurement chosen to suit a particular set of project conditions.

Under perfect conditions this may well be true, although there will always be other factors which affect performance to a lesser or greater extent than such relationships and performances. However, most projects are not carried out under ideal conditions or with a team that will have had any previous opportunity of establishing good working relationships.

Indeed, Liu and Fellows [8] suggest that the presence of many individual organisations within a TMO leads to the '... magnification of organisational complexities...' and that this state then needs to be carefully managed to ensure good project performance.

Projects will always be subject to many uncertainties and risks – even when good relationships exist between members of a TMO – that can be reduced or increased by the use of the correct or incorrect project strategy for dealing with these inherent difficulties. The formulation of the most appropriate strategy is one of the most important tasks that the client has to undertake during the project's life.

The project strategy consists of a number of substrategic areas, which need to be examined in detail by the client so that appropriate decisions can be taken as to which substrategy should be adopted.

Based upon Morris's work [6], these areas should include:

- the roles of the client and third parties;
- the client's project objectives;
- environmental issues;
- quality;
- safety;
- financial objectives, funding and cost planning;
- legal and insurance issues;
- technical and design philosophy;
- project/work breakdown structure;
- milestone schedule;
- risk management;
- project constraints;
- the method of organising the design and construction of the project;
- logistics;
- employment and industrial relations;
- public relations/communications;
- information technology.

Although most of these substrategies will affect, to varying extents, the choice of the method used to procure the project, it is intended to examine only those which are fundamental to that part of the project-implementation process which identifies, examines and eventually selects the most appropriate method of procurement.

These will be discussed in the following chapters and are:

- the roles of the client and third parties;
- the client's project objectives;
- project constraints;
- risk management; and, in much greater detail,
- the method of organising the design and construction and, where appropriate, the funding and operation of the project.

References

1 HM Treasury (1995) *Setting New Standards: A Strategy for Government Procurement*, London: HMSO.
2 Cabinet Office Efficiency Unit (1995) *Construction Procurement by Government: An Efficiency Unit Scrutiny*, London: HMSO.
3 Latham, M. (1994) *Constructing the Team: Joint Review of Procurement and Contractual Arrangements in the United Kingdom Construction Industry*, London: HMSO.

4 Cox, A. and Thompson, I. (1998) *Contracting for Business Success*, London: Thomas Telford.
5 Green, S.D. and Lenard, D. (1999) 'Organising the project procurement process', in *Procurement Systems: A Guide to Best Practice in Construction*, Rowlinson, S.M. and McDermott, P. (eds), London: E & FN Spon.
6 Morris, P.W.G. (1994) *The Management of Projects*, London: Thomas Telford.
7 Walker, D.T.H. (1995) 'The influence of client and project team relationships upon construction procurement performance', *Journal of Construction Procurement* 1, 42–55.
8 Liu, A. and Fellows, R. (1999) 'Culture issues', in *Procurement Systems: A Guide to Best Practice in Construction*, Rowlinson, S.M. and McDermott, P. (eds), London: E & FN Spon.

2   Clients of the construction industry

2.1 Introduction

Project implementation begins with the client, the sponsor of the construction process, who provides the most important perspective on project performance and whose needs must be met by the project team [1]. However, clients of any industry are not a homogeneous group, and it follows that different clients, or categories of clients, will require different and probably discrete solutions to their problems and will present different opportunities.

It is therefore essential that – before addressing the technical, managerial and aesthetic aspects of the project – the identity, nature and characteristics of the client are comprehensively and accurately identified and that the project team is fully aware of, and understands, the client's needs.

The establishment of a definition of the client is essential in order to avoid misunderstandings, and it is proposed that the following meaning will be used throughout this work:

> The organisation, or individual, who commissions the activities necessary to implement and complete a project in order to satisfy its/his needs and then enters into a contract with the commissioned parties.

The latter part of the definition relating to the contractual relationships between the parties is of particular importance in identifying the party legally responsible, under the contract, for carrying out the client's duties, including of course the reimbursement of the commissioned parties.

2.2 The identification and categorisation of clients

The real client

The importance of identifying the real client has been amply demonstrated in both theory and practice [2]. Strange as it may seem, problems often occur as a result of a failure to exercise caution in this respect that present major difficulties for all those involved in the project.

Often, the client organisation consists of numerous disparate, and sometimes competing, groups and factions with their own conflicting agendas and objectives. The increasing use of consortia, partnerships, etc. brought together for the funding, construction and operating of projects can exacerbate the difficulty.

Problems of this sort are most likely to be experienced when dealing with many-faceted organisations in which large numbers of departments and/or individual managers will eventually occupy the facility. In this situation, difficulties will stem from conflicting requirements and standards and it is essential that these are resolved before commissioning any consultants or contractors. The appointment of a single representative to co-ordinate clients' varying requirements will at least reduce the negative consequences of conflicting needs.

In the same way, a situation in which the future occupier of a facility is obtaining funding from an external source, or is a tenant of the client, raises the likelihood of similar conflicts arising.

Establishing the identity of the real client is not always easy, and reconciling the demands of different end-users, both within and outside the client's organisation, always requires considerable expertise, ingenuity, tact and diplomacy. However, the increasing complexity of many clients' organisations, and the complicated structures of most consortia and similar groupings, means that the industry must accept the need to deal with such difficult characteristics as a matter of course.

The categorisation of clients

An examination of past research and literature reveals that the characteristics of clients that are most likely to be relevant to the implementation of construction projects, and more particularly to affect the choice of the most appropriate method of procurement, are:

1   Whether the organisation is publicly or privately owned or funded.
2   The level of knowledge and experience within the organisation in dealing with the construction industry and implementing building projects. This characteristic will also include the degree of expertise within the management of the concern and its policies, philosophy and cultural and other attitudes to project implementation.
3   Whether the project is needed by the client to accommodate his/her own industrial or commercial activities or whether the project is needed to lease, or sell, to others.
4   The activities carried out by the organisation and the resulting project typology.

A detailed examination of these four characteristics reveals a number of key issues:

1   Clients of the construction industry have traditionally been categorised in literature and statistical data as 'public' or 'private' organisations as a reflection of the ownership or source of funding of the establishment.

The characteristics of these two categories of client differ mainly as a result of the source of funding, with publicly financed bodies needing to ensure that the expenditure of taxpayers' money is safeguarded by the adoption of risk-averse and conservative policies. Internal regulations, standing orders and continuous control and auditing of expenditure are also used to ensure public accountability, and each of these safeguards can have the effect of limiting the choice of procurement process.

On the other hand, received wisdom suggests that private organisations are concerned with maximising profits and maintaining share value and dividends for their investors and are therefore prepared to adopt more aggressive policies and take such commercial risks as are necessary to achieve their ends.

In the case of projects implemented by private organisations but funded from external sources, i.e. financial institutions, it is suggested by Turner [3] that more aggressive policies will be adopted than when internal funding is utilised.

Various surveys of client organisations, but particularly those of Hewitt [4] and Masterman [5], both of which were directly concerned with the behaviour of clients when procuring building projects, confirm that this basic difference exists between the public and private sectors.

It is not envisaged that any difficulty will be experienced in identifying into which of these two groups of clients most organisations fall, although the recent increase in the use of private funding for public projects, using techniques such as the private funding initiative (PFI), is now likely to blur the boundaries between the two groups.

2   The level of the client's experience of the construction industry and project implementation has been seen by many researchers, but especially by Morledge [6], as a critical characteristic in terms of client behaviour when dealing with the construction industry.

It has also been amply demonstrated that the client's attitude to all aspects of construction activity is determined by whether they are 'experienced' or 'inexperienced'.

The presence of the following characteristics, in part or total, will determine into which of these two categories the client should be placed.

Experienced clients normally are perceived as exhibiting positive characteristics which include:

• a detailed knowledge and understanding of the construction industry and its procedures;
• a continuing, or regular, involvement with the construction industry, including at times the implementation of large high-value and complex projects;

- the ability to produce a comprehensive initial brief incorporating prioritised objectives for the cost, timing, quality and functionality of the project;
- expertise in the overall management and control of construction projects and of construction consultants and contractors;
- the employment of in-house construction managers and sometimes designers of various construction disciplines;
- a desire to be constructively, consistently and continuously involved during the life of the project without detriment to the powers, responsibilities and status of the appointed consultants and/or contractors.

Inexperienced clients are, on the other hand, seen as exhibiting different, mainly negative, characteristics, which include:

- a lack of continuing or regular involvement with, or knowledge of, the workings of the construction industry or the implementation of construction projects other than some maintenance or minor building works;
- a dearth of expertise in the overall management of construction projects and/or construction consultants or contractors;
- a lack of understanding of the importance of the early production of an initial brief and an inability to produce such a document, or a set of prioritised objectives, without substantial assistance from an external construction consultant;
- a need to make changes to the project parameters throughout the duration of both the design and construction periods and a lack of understanding of, or an unwillingness to accept, the financial and practical consequences of such action;
- a desire to be involved with the project on an inconsistent, random and intermittent basis;
- a weakness for being influenced on construction matters by external parties other than his/her own advisers.

It has been suggested [7] that, in fact, some clients, or their employees, will exhibit a number of characteristics from both of these lists as a result of some consistent but non-continuous involvement with the construction industry.

This experience is most likely to have been obtained when carrying out building works to maintain, expand or upgrade an organisation's primary commercial or industrial activities, and the existence of this category is acknowledged and is referred to in the text as a 'partially experienced' client.

It is also accepted that experienced clients will not always exhibit positive characteristics, and conversely that not all inexperienced clients will exhibit negative behaviour. Indeed, Galbraith [8] suggests that all clients will be influenced more by experience when choosing their procurement strategy than by project-specific factors.

3   The third characteristic relates to the reason for the client's need to implement a building, i.e. whether the facility is required for the housing of his/her own industrial or commercial activities or whether it is required to lease, or sell, to others.

Research [9] has identified these two subcategories as:

- primary clients whose main activity and primary source of income derives from constructing buildings for sale, lease, investment, etc.;
- secondary clients who only require buildings to enable them to house and undertake their own main business activities and whose expenditure on construction represents a small proportion of their annual turnover.

The vast majority of primary clients will, by their very nature, be experienced and largely involved in public or private property development.

Secondary clients will consist of a wide range of large, medium and small manufacturing and service organisations making up the UK's industrial and commercial sectors and will exhibit various levels of experience.

4   The activities of the organisation, i.e. the trade it is engaged in, the business it conducts, the services that it provides and the type of building in which it carries out these activities have been used for many years by observers of the construction industry to categories its clients.

Government statistics have grouped contractors' output figures and other data by the activity carried out within the commissioned facility, for example factories, warehouses, offices, educational and health facilities etc., since the collection of such information began.

This typology has been used as a basis for obtaining information from clients of the construction industry by researchers in an attempt to determine any change in the approach adopted by the various users of the different types of facility when choosing the way in which they procured their project.

Rowlinson and Newcombe [10], for example, found that when examining forty industrial projects the type of production process carried out within the facility had a profound influence on the priorities adopted by the various clients when implementing their projects. On the other hand, Masterman's survey [5] of a similar number of industrial and commercial organisations was only able to identify minor changes in the way the different types of client chose the procurement method.

Despite the lack of firm evidence, it is accepted that there may be a correlation between the typology of the project and the way in which the client chooses what appears to him/her to be the most appropriate project procurement route. However, for the purpose of categorising clients in this work, the typology of clients' activities will be ignored and clients will be grouped on the basis of the first three of the characteristics previously described, that is to say:

- public/experienced/primary clients;
- public/experienced/secondary clients;
- private/experienced/primary clients;
- private/experienced/secondary clients;
- private/partially experienced/secondary clients; and
- private/inexperienced/secondary clients.

A graphic representation of these categories is shown in Figure 2.1.

In terms of the implementation of construction projects, and particularly the selection of procurement methods, it needs to be borne in mind that subcategories of all the listed categories probably exist, but for the purposes of initially establishing the likely characteristics of a client the six suggested classifications should be an adequate guide.

Other methods of classifying clients obviously exist. Most recently, Galbraith [8] has identified groupings of construction industry customers based upon organisational factors which, although reflecting many of the criteria that have been discussed, resulted in the following six classes of client being identified:

- traditional professions, public-sector customer;
- traditional professions, developer/investor customer;
- the 'average' construction industry customer;
- newer professions, developer/investor customer;
- industrial new-build customer; and
- industrial, existing buildings customer.

While it was admitted that the results of Galbraith's research had not yielded a firm method of classifying customers, it was contended that it at least increased the understanding of the need to take account of the client organisations' characteristics and thus their likely approach and attitude to project procurement.

It is acknowledged that no categorisation will ever succeed in totally capturing the many variants and subspecies of its subject. Clients of the construction industry are no different in this respect from any other group of individuals or organisations; some will always exhibit non-stereotypical characteristics and remain unclassified. In the main, however, it is suggested that the majority will remain remarkably faithful to the

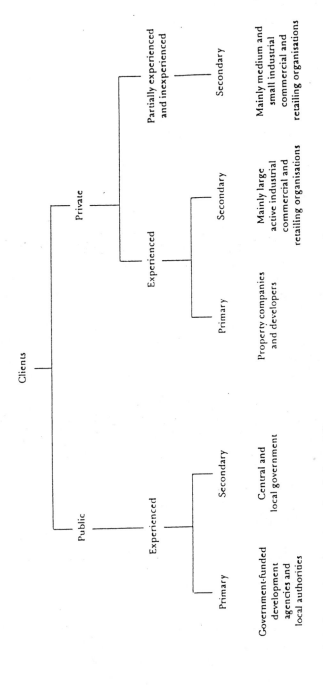

Figure 2.1 Categories of construction industry clients

attributes that have been identified, although care must always be taken to ensure that the specific characteristics of each organisation or individual are established as a first priority.

2.3 Clients' objectives and needs

The Latham Report [1] suggested that a client's project needs are:

- obtaining value for money;
- ensuring the project is delivered on time;
- having satisfactory durability;
- incurring reasonable running costs;
- being fit for its purpose;
- being free from defects on completion;
- having an aesthetically pleasing appearance;
- being supported by meaningful guarantees.

There is little doubt that these requirements reflect the needs of most clients, either in whole or part, but an attempt will now be made to establish what are the client's needs and objectives that are particularly related to the project procurement process, and perhaps more contentiously to prioritise them for different categories of client.

No matter what the client's needs and objectives are, it is essential in exactly the same way that the client's characteristics must be identified and understood, that the project team establishes and understands all of the client's requirements as accurately and as quickly as is appropriate and possible.

Apart from the Latham Report, many detailed studies of the needs and objectives of clients have been carried out over the past three decades, and the most relevant of these are now examined.

In 1975, the Wood Report [11] revealed the results of case studies of fifty building and civil engineering projects which showed that clients consistently mentioned the need to meet the criteria of cost, low maintenance charges, time, quality, functionality and aesthetics as being necessary for a project to be considered successful.

Typical clients' requirements were defined by Bennett and Flanagan [12] and can be summarised, but not prioritised, as:

- a functional building, at the right price;
- quality, at the right price;
- speedy construction;
- a balance between capital expenditure and long-term ownership costs;
- identification of risks and uncertainty associated with the project;
- accountability, particularly in the public sector;
- innovative design/high-technology buildings;

- maximisation of taxation benefits;
- flexibility to enable design to be changed;
- a building which reflects the client's activities and image;
- minimisation of future maintenance;
- the ability to keep any existing buildings operational during the construction period;
- an involvement in, and a need to be kept informed about, the project throughout its life.

It was not suggested then, and it is not proposed now, that this list is exhaustive or that the typical client, if such exists, would aim to, or be able to, satisfy all these needs in one project, at least to the same level of satisfaction. Rather, it is suggested that the list reflects most of the requirements of most clients.

In the same year that this list was produced, the landmark report *Faster Building for Industry* [13] was published. The results were based upon research that had been carried out in the main to establish clients' attitudes to the timing and speed of their projects, and concluded that few clients were interested in speed for its own sake.

The report also found that very little consideration had been given to the influence of time on the financial aspects of projects, with clients, as a general rule, not being prepared to incur additional costs in order to achieve fast construction times and being uncertain about the role of time in the construction process.

Hewitt [4], in a survey of twenty-one public and private client organisations carried out in 1985, identified the 'real needs' of clients as:

- Certainty of cost and time and a reduction of client vulnerability to unanticipated extra costs and time overruns.
- Private sector clients required contractual arrangements which provided them with the flexibility to change the design during construction.
- A strong desire to be actively involved and kept informed throughout the design and construction period.
- Consultants should be more forthcoming with positive and constructive advice and more prepared to recommend new procurement methods.
- Consultants should be prepared to accept legally binding predetermined construction programmes and to have their fees calculated on a non-adjustable lump sum basis.
- The industry should tailor its products to suit the demands of its clients.

In 1988, *Faster Building for Commerce* [14] was published, drawing on sixty case studies, detailed information from 260 other projects and a tatistical analysis of 8,000 projects. This report established that clients' priorities were determined by their need to minimise the commercial risks associated with property development and that their main concern was the functionality of their buildings.

Fast and punctual construction periods were demanded as a matter of course, and quality standards were high and often very demanding. In addition, the facility to revise cost targets and to introduce changes to their schemes at all stages of the project was a prime requirement.

In the same year, the Centre for Strategic Studies in Construction, based at the University of Reading, produced a report [15] on the future of the building industry in the UK which maintained that although the modern client is more sophisticated than in the past his/her needs, in terms of construction projects, broadly remained the time-honoured triumvirate of time, cost and quality.

Both of these last two reports confirmed that clients wanted certainty of performance in all three of these criteria, did not want any surprises during the implementation of their projects and specifically required:

- value for money;
- a durable and easily maintained building with affordable running costs;
- no latent defects and rapid rectification of any minor problems that may occur;
- clear allocation of responsibilities among the members of the project team;
- a minimal exposure to risk;
- an early indication of a firm price for the project;
- comprehensive information on any future contractual claims;
- an early start on construction work with minimal interference from external sources such as planning, building control;
- a non-confrontational business relationship with the contractor, who should give guarantees and good 'after sales' service.

While these reports and studies identified the client's basic needs and primary objectives, and in some cases examined these requirements in more detail and determined the client's criteria for success, there had been little attempt to prioritise these needs or to examine the individual requirements of different categories of client.

Masterman's 1994 study [16] attempted to address this oversight by surveying more than sixty client organisations and asking them to rank a list of criteria, which had been established from previous research, that they considered essential to ensure project success on their last, or only, project.

The results of this exercise are shown in Table 2.1. A statistical analysis carried out on the resulting data confirmed that the ranking of the listed criteria was meaningful, but that the differences in the ranking of the needs by the three main categories of client were not significant.

Table 2.1 Mean rating, by all categories of client, of objectives/needs

Clients' objectives/needs	Public experienced primary clients		Public experienced secondary clients		Private experienced primary clients		Private experienced secondary clients		Private partially experienced secondary clients		Private inexperienced secondary clients	
	Mean	Rank	Mean	Rank	Mean	Rank	Mean	Rank	Mean	Rank	Mean	Rank
Certainty of final cost	5.00	=1	5.00	=1	4.00	=4	3.50	=6	4.35	3	4.62	1
Accountability	5.00	=1	4.00	6	3.00	=10	4.45	1	3.69	7	3.00	10
Value for money	4.50	3	5.00	=1	4.00	=4	4.17	=2	4.18	4	3.59	=7
Lowest possible tender	4.25	=4	5.00	=1	5.00	=1	3.55	5	3.75	6	3.94	4
Actively involved and informed	4.25	=4	5.00	=1	4.00	=4	4.17	=2	4.83	1	4.27	3
Single point of contact	4.25	=4	3.00	=7	3.00	=10	3.00	10	3.54	8	3.59	=7
Elimination of/minimise risk	4.25	=4	3.00	=7	4.00	=4	3.08	9	4.14	5	3.75	6
Minimum design and construction period	3.75	=8	3.00	=7	5.00	=1	3.50	=6	3.46	9	3.88	5
Certainty of completion date	3.75	=8	5.00	=1	5.00	=1	4.17	=2	4.46	2	4.35	2
High-quality and innovative architecture	3.25	10	2.00	11	3.50	=8	2.67	11	3.15	11	3.13	9
Ability to change design	2.75	11	3.00	=7	3.50	=8	3.17	8	3.29	10	2.89	11

Only thirty-seven of the sixty-three participants in the survey responded to the request to rank the listed criteria, and nearly all (96 per cent) of those that did not participate were partially experienced or inexperienced clients. Thus, the less experienced participants appeared to be unable, or unwilling, to identify and/or prioritise their needs even when provided with a predetermined list of criteria.

What conclusions can be drawn from all this evidence?

There is little reason to doubt that functionality/quality, cost and time remain clients' primary objectives, although the specific definition of these criteria will vary from client to client and project to project, e.g. time could refer to the need for a rapid start on site or to an accelerated finish or a minimal construction period.

It has long been recognised that these three primary objectives are inter-related and conflicting and that it is impractical to try to achieve all three at the same time on the same project. At least one – more probably two – will need to be sacrificed to some extent, and individual clients will need to weight each of the criteria to suit their own organisation's particular circumstances and the project's technical, commercial and other characteristics [17].

The weight given to each objective will obviously vary, although the balance that was initially proposed may not always be achievable and a compromise between conflicting criteria may have to be reached. Notwithstanding any such difficulties, the client's needs must always be accurately and clearly determined.

Achieving adequate functional performance and the right level of quality is usually the dominant primary objective in most construction projects, and the client needs to establish a minimum acceptable level of functional performance and quality for his/her project at an early stage and incorporate his/her requirements within the project brief in sufficient detail to avoid misunderstandings.

The primary objective of time needs to be considerably refined before it can become meaningful in the context of the brief as it can refer to any number of variants of this basic need.

In the same way, an acceptable level of cost – the third primary objective – needs to be established for every project, and again different variants of this need will apply to different clients and specific projects.

Once the dimensions of the three basic objectives have been determined, weighted and a compromise reached between them, there may well be a need to take into account the secondary objectives that clients often set for their construction projects.

These secondary objectives include:

- the need to use members of the client's own specialist staff, own labour force or the services offered by a subsidiary or sister company of the client's organisation;

- the provision of training for the client's workforce that will eventually occupy and operate the new facility;
- the incorporation of the client's existing plant or structures within the new facility;
- the transfer of all possible risk to parties other than the client;
- the incorporation, within the design and/or construction periods for the building, of separate parallel operations being carried out by designers and/or contractors employed directly by the client on the same site;
- the minimisation of maintenance and running costs;
- the client's need for single-point responsibility, the appointment of specific design consultants or the novation of the design team;
- the need for public accountability;
- the need for flexibility within the implementation of the project to enable changes to be made to the design;
- maintaining the operational viability of sections of an existing facility during the construction of extensions or the carrying out of alterations.

Although such objectives are termed secondary, they can, particularly in combination, have considerable influence on the selection of the procurement method used for the project.

Each client will obviously have different specific primary and secondary objectives for each of his/her projects, but it is suggested that the list of needs shown in Table 2.1 represents the requirements that should be and usually are, consciously or subconsciously, considered by clients when establishing their objectives. It is these objectives that are the key determinants of procurement method suitability [18].

2.4 Project constraints and risk

This section is only intended to provide the reader with an overview of the constraints that can be experienced, and the risks that are inherent, in implementing construction projects. Specific constraints and risks associated with the procurement process are described within the chapters relating to specific procurement system categories.

Constraints

Most construction projects will exhibit characteristics which can be classified as constraints and which need to be identified so that they can be considered when formulating the project strategy.

The effect of these constraints is to influence or restrict actions which might otherwise assist in the successful achievement of the client's objectives. Often, the constraints will be physical, for example difficult ground conditions, high water tables, planning authority requirements, but they can also be the consequence of the client organisation's policies, culture or internal regulations.

The constraints stemming from internal rules restrict the actions of managers in order to safeguard the interests of shareholders or taxpayers. For example, the public client may only be able to accept tenders which enable him/her to enter into a type of contract which ensures that a lump sum price is agreed with the most acceptable bidder and that it is fixed for the duration of the project. Such an approach would eliminate a substantial number of procurement methods which would otherwise be available.

On the other hand, the private client may be able to enter into financially open-ended procurement arrangements but may need to complete the project within a specific time-scale in order to ensure that the new product or service will be available ahead of a competitor in the market-place.

When initially examining the project, all of its discrete constraining factors need to be identified. At the same time, all of those parties within the client's organisation having an interest in the development, together with any external source which may be able to exert any restraining influence, need to be considered.

These factors need to be assessed, if necessary using one of the number of techniques now available for this purpose, in order to quantify their actual constraining effect on the project objectives and the procurement process.

Risk management

The management of risk is a wide-ranging and well-documented subject which is of vital importance in ensuring the successful management of the project-implementation process and particularly the selection of the most appropriate building procurement method, each of which carries a different level of risk for each of the main participants.

Although it is not the purpose here to provide a guide to risk management, a brief summary of the principles of this area of the project strategy will enable the reader to understand its relation to the procurement process.

Smith [19] points out the theoretical difference between risk and uncertainty, but suggests that the terms are, for all practical purposes, interchangeable. An acceptable definition might therefore be that risk is present when there is more than one possible outcome of a decision or action, with the probability of one or more outcomes being unknown.

Perry [20] suggests that the process of risk management consists of four stages:

1 Identification of the sources of potential risk.
2 Assessment of their effects by means of risk analysis.
3 Development of the method of responding to the identified risks.
4 Providing for the residual risk.

The sources of risk when implementing construction projects include:

- the client;
- government regulatory authorities;
- project funding and other financial matters;
- design;
- local conditions;
- construction contractors and their activities;
- logistics;
- estimating data;
- inflation;
- exchange rates;
- *force majeure.*

When endeavouring to identify the sources of risk, the discrete characteristics of the project, which are likely to be the main sources of risk, should be examined to determine which of these might endanger a successful outcome to the project.

The effects of the risks, once they have been identified, need to be quantified by means of risk analysis, with the choice of the analysis technique – many of which can be extremely complex – being determined by the level of expertise and experience of the project team and the size and complexity of the project.

In its crudest form, risk analysis will treat each major risk as a single issue and assume no interdependence with others or attempt to quantify the probability of the risk actually occurring; more sophisticated techniques take into account probability and interdependence.

Risk response should be thought of in terms of *avoidance, reduction, transfer* or *retaining.*

The first two methods will be used where the identification and analysis of the project risks indicate a need for, say, amendments to the physical scope or nature of the project, further site investigations, redesign of specific elements or a different construction methodology in order to achieve a reduction or avoidance of the identified risks.

The transfer of risk requires that a party other than the client takes responsibility for part, or all, of the consequences of the identified risk should it actually occur. Such an action demands that the risk is accurately and comprehensively identified, that the parties which assume the risk have the necessary ability to control and deal with the consequences and that the transfer is in the best interest of the client, who will usually pay a premium for the transfer.

Retaining the risk is, in most cases, likely to be the last resort as many clients are risk averse and keen to allocate any potential problems to others. Where responsibility is retained, an allowance is usually included within the project budget as a contingency fund. Insurance cover for well-defined specific risks can be obtained, but in most cases it is unlikely to be cost-effective. Whichever of these four routes are taken, the principle of allocating the risk to the party most capable of dealing with it at the minimum cost to the project should always be followed.

Four of the five elements of the project strategy that are considered to be fundamental to the assessment and selection of the most appropriate method of procurement have now been examined in sufficient detail to enable the reader to have an understanding of the role of the client, the client's needs and objectives, the role of project constraints and the basic principles of risk management.

The remainder of this work is now devoted to examining the methods of procuring the project and the means by which such methods are, and should be, chosen.

References

1   Latham, M. (1994) *Constructing the Team: Joint Review of Procurement and Contractual Arrangements in the United Kingdom Construction Industry*, London: HMSO.
2   Cherns, A.B. and Bryant, D.T. (1984) 'Studying the client's role in construction management', *Construction Management and Economics* 2, 177–184.
3   Turner, A. (1997) *Building Procurement*, 2nd edn, Basingstoke: Macmillan.
4   Hewitt, R.A. (1985) 'The procurement of buildings: proposals to improve the performance of the industry', unpublished project report submitted to the College of Estate Management for the RICS Diploma in Project Management.
5   Masterman, J.W.E. (1989) 'The procurement systems used for the implementation of commercial and industrial building projects', unpublished MSc thesis, University of Manchester Institute of Science and Technology.
6   Morledge, R. (1987) 'The effective choice of building procurement method', *Chartered Surveyor* July, 26.
7   Gameson, R.N. (1992) 'An investigation into the interaction between building clients and construction professionals', unpublished PhD thesis, University of Reading.
8   Galbraith, P.J. (1995) 'The development of a classification system for construction industry customers', EPSRC research report, Department of Construction Management and Engineering, University of Reading.
9   Naphiet, H. and Naphiet, J. (1985) 'A comparison of contractual arrangements for building projects', *Construction Management and Economics* 3, 217–231.
10 Rowlinson, S. and Newcombe, R. (1986) 'Design–construction organisation', paper presented at the International Association for Bridge and Structural Engineering Conference, Zurich, Switzerland.
11 Building Economic Development Committee (1975) *The Public Client and the Construction Industries (The Wood Report)*, London: National Economic Development Office.
12 Bennett, J. and Flanagan, R. (1983) 'For the good of the client', *Building* 27, 26–27.
13 Building Economic Development Committee (1983) *Faster Building for Industry*, London: National Economic Development Office.
14 Building Economic Development Committee (1988) *Faster Building for Commerce*, London: National Economic Development Office.

15 Centre for Strategic Studies in Construction (1988) *Building Britain 2000*, Reading: University of Reading.

16 Masterman, J.W.E. (1994) 'A study of the bases upon which clients of the construction industry choose their building procurement systems', unpublished PhD thesis, University of Manchester.

17 Walker, A. (1996) *Project Management in Construction*, 3rd edn, London: Collins.

18 Jennings, I. and Kenley, R. (1996) 'The social factor of project organisation', paper presented at the CIB W92 (International Commission on Building) Procurement Systems Symposium, University of Natal, Durban, South Africa.

19 Smith, N.J. (1999) *Managing Risk in Construction Projects*, Oxford: Blackwell Science.

20 Perry, J.G. (1987) 'Risk management', course notes, MSc in Construction Management, University of Manchester Institute of Science and Technology.

3 The concept and evolution of building procurement systems

3.1 Introduction

Once the roles of the client and any external parties have been identified, the client's needs and objectives established, any constraints examined and quantified and all risks identified, assessed and allocated to the parties best placed to deal with them, the selection of the most appropriate procurement process must be considered.

The definition of the procurement process developed by the International Commission on Building (CIB W92) during its 1997 meeting was:

> ... a strategy to satisfy client's development and/or operational needs with respect to the provision of constructed facilities for a discrete life cycle.

The element of the strategy that will be examined in this work commences once the client has made the critical decision to build and ends when the most appropriate method of procuring the project has been chosen.

This method will be referred to throughout as the *procurement system*, and it is only right therefore that a little time is spent in establishing a definition of this term.

McDermott [1] maintains that this description should encompass not only the method used to design and construct the project but also the cultural, managerial, economic, environmental and political issues raised by the implementation of the procurement process.

While the importance of considering these aspects of the process should not be underrated, the author is of the opinion that they should be treated as described in Chapter One, i.e. as a subelement of the project strategy rather than as an integral part of the procurement system itself.

This approach is defended not only on the basis of the general acceptance and well-established use of the term within the industry specifically to describe the various methods of implementing projects but also because the subelements do not have the ability to change the procurement systems themselves but rather affect the way in which they should be selected and used.

It is also felt that, provided all of the subelements are considered when formulating the project strategy, it is somewhat academic as to whether the commonly accepted definition of the term should be altered or widened. Having said this, there is a need to accept that contemporary procurement systems can now embrace not only the design and construction of projects but also their financing, operating, facilities management, etc., and it is therefore proposed that for the purposes of this work the following definition will be used:

> A procurement system is the organisational structure adopted by the client for the implementation, and at times eventual operation, of a project.

The choice of building procurement systems available to clients is now so wide that the need to carry out the selection of the most appropriate method in a disciplined and objective manner should be self-evident. However, as we will see later, when the way clients actually carry out this process is examined, selection is often carried out in an haphazard manner. As a first step in assisting clients to overcome this problem, it is suggested that it would be helpful if the main procurement systems were categorised.

3.2 Categorisation of procurement systems

There are a number of ways in which categorisation can be achieved, for example:

1   By the amount of risk taken by each of the participating parties – construction projects carry varying degrees of risk, some covered by statutory liability, some covered by insurance and the remainder, often described as speculative risk, apportioned within the precontract documentation to the most appropriate party [2].

    However, categorisation by the degree of risk, for example high, medium or low, does little to inform the decision-maker of the fundamental differences between the various systems.

2   By the level of information available or required at the time construction contracts are let – the amount of overlap required between completion of the design and the commencement of construction has always been of interest to clients, particularly those that are impatient to commence work on site, as it has a direct relationship to the speed with which the project can be physically started.

    However, any categorisation using this criterion is likely to be one-dimensional and misleading to the decision-maker.

3   On the basis of the way in which the contractor is reimbursed – as many procurement systems allow for reimbursement to be made in the same way, this method would not assist in identifying the individual

systems and is therefore considered to be an invalid means of categorisation.

4   By the way in which the interaction between the design and construction, and sometimes the funding and operation, of the project is managed – this method enables the fundamental issues of the relationship between the main elements of the project to be identified from the categorisation and at the same time ensures recognition of those procurement systems that are contained within each category.

This last method, which is based upon an original approach devised by Perry [3], is considered to be, for the purpose of assisting in the selection of the most suitable procurement system, the most appropriate means of classification.

Therefore, for the purposes of this work, the following categories of building procurement systems have been adopted:

1   Separated procurement systems – where the main elements of the project-implementation process, i.e. design and construction, are the responsibility of separate organisations, e.g. design consultants, quantity surveyor, contractor. The client has all of the members of the project team to deal with and is responsible for the funding and eventual operation of the facility.

     This category contains the *conventional system*.

2   Integrated procurement systems – where one organisation, usually but not exclusively a contractor, takes responsibility for the design and construction of the project and, in theory at least, the client only deals with one organisation.

     Design and build, novated design and build, develop and construct, the *package deal* method and the *turnkey* approach are the main systems in this group. In the case of the last method, the contractor may well provide or arrange funding for the project and be responsible for the subsequent operation of the facility.

3   Management-orientated procurement systems – where the management of the project is carried out by an organisation working with the designer and other consultants to produce the design and manage the physical operations which are carried out by works, or package, contractors.

     When using systems within this category, the client will need to have a greater involvement with the project than when employing any of the other methods described in the previous two categories.

     The main systems contained in this group are *management contracting, construction management* and *design and manage*.

4   Discretionary systems – where the client lays down a framework for the overall administration of the project within which he/she has the discretion to use the most appropriate of all the procurement systems contained within the other three categories.

Partnering and the little used *British Property Federation system* are the two main frameworks included within this grouping.

Figure 3.1 illustrates the suggested procurement system categorisation.

It should be said that although categorisation is important in the context of establishing a discipline for examining and selecting procurement systems, it is accepted that its use is likely to be restricted to these activities and that it is questionable as to whether it will be, or needs to be, more widely applied.

These groupings and the individual systems themselves are discussed in the following chapters, but it should be understood that whereas, in the majority of projects, the use of one procurement system will normally ensure that the client's needs are met on larger and more complex projects it may be necessary for several of the methods to be used in combination, or singly in different geographical sections of the same scheme, and it is not unknown for a bespoke procurement system to be designed for a specific project.

3.3 The evolution of contemporary procurement systems

Introduction

Having established the concept of procurement systems, the evolution and level of use of contemporary procurement methods over the past half century or more is now examined in some detail using the numerous official and semi-official reports that were produced during this period.

The vast majority of the construction projects before the Second World War (1939–1945) were implemented by conventional methods of procurement that had remained virtually unchanged for over 150 years. Since that time, however, the number of different procurement systems available has substantially increased, often as a result of importation from the USA and perhaps, more significantly, as a result of the willingness, generally because of frustration at the construction industry's poor performance, of an increasing number of clients to sponsor and use new methods.

Four phases in the development of contemporary procurement systems can be broadly identified. The first was a period of sustained economic growth when the use of conventional methods of procurement still prevailed. The second was a period of recession characterised by a relatively modest increase in the use of non-conventional procurement systems. The third phase was a time of post-recession recovery during which the most experienced clients introduced new procurement methods and design and build and management-orientated systems substantially increased their share of the available workload.

The fourth and final period covers the last decade and is ongoing. It has so far contained nearly equal periods of recession and recovery as well as the advent of partnering, an increase in the use of the various forms of the private finance initiative, and efforts by government to improve the industry's

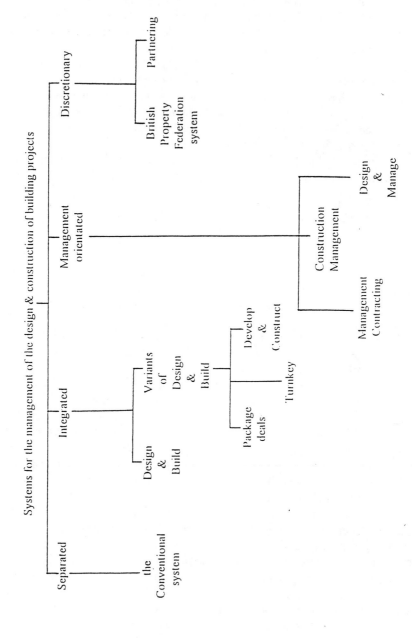

Figure 3.1 Categorisation of building procurement systems

performance through the publication of two major reports and changes in the way it manages its own projects.

Chronologically, these periods relate to the years 1945–1972, 1973–1980, 1981–1990 and 1991 to the present time – over half a century during which the changed attitudes and needs of client organisations have done more than any other factor to increase substantially the number and types of available procurement systems.

1945–1972

After the end of hostilities, the demand placed upon the building industry rapidly increased in terms of both workload and complexity. Despite this, the way in which projects were organised remained largely unaltered, particularly in the public sector, where the majority of the work was still being let on the basis of open competitive tendering despite the Simon Report [4] of 1944 having strongly recommended the use of selective bidding.

The Phillips Report [5] published in 1950 reiterated this recommendation and, in addition, highlighted the need for greater co-operation between all of the parties involved in the construction process. By now, however, some innovative procurement systems, such as negotiated tenders and design and build, had begun to be used on a very limited scale by the private sector and central government.

Criticism of the lack of sound relationships and co-operation between the members of the project team and their mutual clients was contained within the 1962 Emmerson Report [6], which made the now well-known comment that:

> In no other important industry is the responsibility for design so far removed from the responsibility for production.

Emmerson came to the conclusion that there was still a general failure to adopt enlightened methods of tendering despite the recommendations of earlier reports, but also noted the growth of package deal and 'other forms' of placing and managing contracts. No examples of such forms, other than serial tendering, and a passing reference to the CLASP (Consortium of Local Authorities Special Programme) system of industrialised building were provided.

In 1964, the Banwell Report [7] was published and expressed concern at the failure of the industry and its professions to think and act together or to reform its approach to the organisation of construction projects. The report also reiterated the recommendations of the Simon Report [4] and other previous committees and working parties in that the use of selective tendering should be more universally applied together with the use of non-conventional methods of procurement, such as negotiated and serial tendering, where appropriate.

The 1967 review of the Banwell Report, *Action on the Banwell Report* [8], found that some progress had been made since the 1964 report on the preplanning of projects, although the professionals had done little to de-restrict their practices. An increase in the use of selective tendering was noted and the industry was again urged to increase the use of serial and negotiated tendering. It was also noted that a number of guides published over the decade before 1965, and intended to assist clients in organising and planning their construction projects, had not been sufficiently publicised and circulated, with the result that they had been under used.

Higgins and Jessop [9], in a pilot study which examined communications in the building industry and which was sponsored by the National Joint Consultative Committee, were probably the first to suggest that overall co-ordination of design and construction should be exercised by a single person or organisation, although it was some years later before such a philosophy gained general acceptance.

The early to mid-1960s were a time of economic expansion, rapidly developing technology, changing social attitudes, the demand for more complex and sophisticated buildings and, not least, the client's increased need for faster completion at minimum cost. All of these factors generated considerable activity within the industry, as a consequence of which the general standard of performance and organisation improved.

The need for cost-effective faster completion of projects stemmed from the revived activities of property developers following the boom in urban development and increased activity within the industrial sector. Developers and industrial organisations were unhampered by the standing orders or restrictive procedures necessarily adopted by the public sector and were more open to suggestions for the use of, what were then, unorthodox arrangements for the provision of their building projects. Negotiated contracts and package deals were frequently entered into by the private sector, and there was much discussion, but much less real progress, about the early involvement of contractors in project planning in order to benefit from their practical expertise.

Overall, this period from the end of the war to the early 1970s was a time of sustained and almost uninterrupted economic growth, during which, in terms of construction procurement, conventional methods prevailed with only a relatively small number of projects being carried out using non-conventional procedures. This was despite a proliferation of reports recommending their use and advocating the adoption of a more co-operative approach by all members of the project team.

1973–1979

This second phase was a period of recession, if not depression and instability, and commenced as a result of unexpected and large price increases in crude oil imposed by the oil-producing countries coupled with high inflation caused by the previous economic boom. This stage is seen to last until the end of 1979, although the effect of the oil price increases on the economies of Western countries were still being felt during the early 1980s.

During this period, government-sponsored studies of the industry tended to be related more specifically to individual sectors of the industry rather than to the more generalised investigations that were carried out in the 1950s and 1960s.

Of the number of studies emanating from the National Economic Development Office (NEDO), the Wood Report [10] was the only one specifically to examine purchasing policies and procurement practices, although even then the examination was restricted to the public sector. The report found that the procurement systems used by public authorities were inappropriate to the circumstances of the projects surveyed, although the use of non-conventional methods of procuring construction work amounted to 40 per cent of the projects examined.

Various official and unofficial reports were produced during this phase which drew unflattering comparisons between the performances of the British and a number of foreign construction industries. The two case studies carried out by Slough Estates in 1976 [11] and 1979 [12] were particularly damning.

The 1976 study found that the overall time taken to implement an industrial project in the UK was considerably longer than in other countries, and the eventual cost of developments in the UK was considerably higher than all but one of the other seven countries surveyed. The reasons for this poor comparative performance were considered to stem from the unnecessarily lengthy and complex design and pricing process and the time taken to obtain the necessary statutory permits which, when related to the high level of interest rates, inflation and prolonged design and construction periods, led to high levels of cost.

The 1979 report endorsed the findings of the first report, but conceded that there had been a general improvement in the intervening 3 years which was attributable to the effects of the recession and the resulting low level of building activity.

The recommendations made by the two studies in the context of the procurement process urged the simplification of design and construction procedures, the improvement of the management of construction, the establishment by clients of their real needs and more effective briefing of the project team.

These two reports were probably the first formalised examples of a trend, which emerged during the late 1970s and has continued since, of some major client organisations forcibly voicing their dissatisfaction with the performance of the construction industry.

In 1978, the Building and Civil Engineering Development Committees combined to produce *Construction for Industrial Recovery* [13], a study which sought, among other aims, to obtain the views of industrialists on the adequacy of the products and services offered by the construction industry.

The report, based upon a survey of 500 firms and thirty-two case studies, concluded that industrial firms carried out their construction projects using a variety of procurement methods, with many choosing the traditional route. It was also established that the average industrial user was not aware of the complexities of the construction process, or more particularly the alternative methods of acquiring buildings, and that contract procedures often failed to meet the needs of the manufacturing sector of industry.

The Royal Institution of Chartered Surveyors published a report [14] in 1979 which was based upon a study of the contribution made by different design and contract procedures to a project's cost and time performance, especially in the UK and US construction industries. The document contained a number of conclusions which are relevant to the procurement process:

1   A very wide range of procurement systems is used in the private sector in the USA compared with a very narrow range in the UK, where central and local government is the dominant influence and public accountability, rather than economy, is essential.

2   Major time and cost penalties are likely to be incurred if detailed design is divorced from construction.

3   The use of construction management in the USA has grown as a result of this system being capable of giving the design team control over the construction time and cost.

4   The range and variety of procurement systems has proliferated on both sides of the Atlantic because of escalating costs, increasingly complex designs, the increased size of projects and the more onerous demands of owners.

5   Clients in both countries were found to be dissatisfied with conventional procedures. In the USA, this dissatisfaction results from the increase in claims, and subsequent litigation, as well as a lack of cost control during the design stage.

6   Clients in the UK are more conservative than their US counterparts, who are prepared to experiment with the whole range of procurement methods, particularly – at the time of the survey – construction management.

The theme of most of the reports published in the 1970s reflected this conservatism as a diminishing number of clients were prepared to commit themselves to implementing projects in an uncertain economic climate. There was also increasing concern among those clients who continued to carry out construction works at rapidly increasing material and labour costs, high inflation and falling demand for their products, all of which were made worse by the delays, overruns and other difficulties associated with the UK construction industry.

Within this second phase, the use of conventional procurement methods still accounted for much of the construction work, although the use of management contracting, and to a lesser extent design and build, continued to increase.

1980–1989

The third of the chronological phases started in 1980 and finished about a decade later. It has been described as a post-recession period of adjustment and recovery, during which changes took place as a result of long-term shifts in the structure of the industry, such as the emergence of labour-only subcontracting, and changes in the attitudes of major clients.

This last characteristic is best reflected in the launching, at the end of 1983, of the British Property Federation's (BPF) *System for Building Design and Construction* [15]. This body, which represents the majority of UK property development organisations and a number of large retailers and commercial companies, had concluded that many existing procurement systems caused delays, were inefficient, could increase costs and could cause and sustain confrontational attitudes between consultants and contractors which were contrary to the best interests of the client.

The Federation had set up a working party, assisted by a number of consultants, to draft 'an improved management system for the building process appropriate to members of the BPF' and eventually produced its own system of procuring and implementing a building project. The system reflected US practice and the experience gained by its members in using all of the various existing procurement methods that had become established over many years. However, although the new system attracted much comment at the time, there is no evidence to suggest that this new approach captured the imagination of many clients or a significant percentage of the available workload.

Two important government-sponsored reports [16,17] were published in 1983 and 1988. Both dealt with the widely held belief that the process of procuring new industrial and commercial buildings was unnecessarily long and difficult. This view had already been expressed in the reports of the 1970s, which compared UK performance unfavourably with that of the building industry's performance in a number of overseas countries.

The 1983 study found, in relation to the procurement methods used on industrial projects, that whereas conventional methods of procurement could give good results the use of non-conventional techniques tended to result in faster progress, although it was also established that the use of approximate bills of quantity and negotiated tendering led to faster project implementation. It was also determined that over half of the projects surveyed were carried out by conventional means, about one-third by the design and build method and the remainder by some form of management or other approach.

The 1988 report, which dealt specifically with commercial projects, identified considerable variability in time performance even on similar projects. Project outcomes had been determined not only by the form of organisation but also by the early development of a comprehensive project strategy and timetable. Two-thirds of the projects surveyed had been carried out using conventional methods, with the remainder being more or less evenly divided between design and build and management methods.

Both studies found that about one-third of all industrial and commercial projects were completed on time but that the remainder overran their planned times by one month or more, thereby confirming the concern expressed by many clients during this period at the industry's inability to satisfy their basic needs.

The frustration of clients at this inability resulted in the British Property Federation's production of its own system of managing projects. Other major clients, mainly those implementing megaprojects in London and the south-east of England, began in the mid-1980s to formulate individually their own forms of procurement in order to satisfy more efficiently their own particular objectives and interests.

The emergence of this type of expert private-sector client, who has the necessary in-house resources to manage large projects and a substantial ongoing construction programme, is one of the phenomena of the period, with the demand side of the industry despairing of the supply side ever putting its house in order. Clients had thus identified the need to develop bespoke methods of procurement, mainly based on the use of construction-management techniques, in order to ensure that their needs were met effectively.

In this third phase, conventional methods remained the most widely used techniques, although there was a substantial increase in the use of design and build and a continued use of various forms of management approaches. There was some evidence, however, which suggested a reduction in the use of management contracting during the latter years of the period, which resulted from some clients' dissatisfaction with the performance of this method on large projects.

1990 to the present time

The fourth phase started around 1990, although the trigger for the recession that began to overtake the industry at about this time had its roots in the tightening of government monetary policy in 1988 and 1989 and the consequent withdrawal of property developers and industrialists from the capital development market.

The first half of the decade, in particular, was a time of low economic growth and uncertainty in business and finance with social pressures mounting and environmental issues dampening enthusiasm for some major projects. Government capital spending, which peaked in the 1980s, was cut back and some 500,000 construction-related jobs were lost during the decade and more than 16,000 construction companies became insolvent [18].

Even in 1997, when there were some real signs of recovery and orders had been increasing since the mid-1990s, the annual output of the industry was still some 20 per cent below the 1990 level, and it was not until the very last years of the century that it could be said with any certainty that recovery had been achieved.

It was against this somewhat volatile background that the two defining construction industry reports of the decade were produced. In 1994, 'Constructing the Team [19], a government- and industry-sponsored report describing a wide-ranging review of the industry carried out by Sir Michael Latham, was published, and 4 years later the findings of the government's construction task force led by Sir John Egan was made public in *Rethinking Construction* [20].

The Latham report was jointly commissioned by government and the industry, with the participation of many major clients, 'to consider procurement and contractual arrangements and the current roles, responsibilities and performance of the participants, including the client'. It was a report produced by a single individual, not a working party or committee; in the author's own words 'It is a personal report of an independent, but friendly, observer'.

In essence, the review endeavoured to put forward solutions to problems it had identified which were inhibiting clients from obtaining the high-quality projects they required. The report concluded that an enhanced performance could only be achieved by team work in an atmosphere of fairness to all of the participants – the process of finding 'win–win' solutions.

Altogether, some thirty recommendations were made, ranging from the need for government to commit itself to becoming a best-practice client to the proposed introduction of a Construction Contracts Bill to give statutory backing to the recently amended Standard Forms of contract.

Specific recommendations relating to the procurement process included:

1   The publication by government of a Contract Strategy Code of Practice, which, in terms that would enable it to be used by inexperienced as well as all other clients, should provide guidance on briefing, the formulation of a project strategy and procurement.
2   The production of a family of contract documents based upon the New Engineering Contract.
3   The setting up of registers of consultants and contractors who have been approved to carry out certain types and values of projects.
4   Tenders should be obtained by public and private clients in accordance with European Union Directives and the National Joint Consultative Committee for Building (NJCCB) code of procedure. Design and build proposals should only be obtained using proposed procedures that are fair and reasonable to all involved parties. All bids should be judged upon the basis of the client obtaining best value for money.

5   The use of partnering arrangements, subject to an initial competitive tendering process, should be encouraged in the public sector.

Many other proposals, not directly related to procurement and covering training, dispute resolution, payment, design, teamwork on site, etc., were contained within the review together with a programme of implementation involving both government and bodies representing the industry and clients. Although initially accepted in principle by all those involved, putting the numerous recommendations into practice has been slower than envisaged and work particularly within government is still under way, although somewhat overtaken by the recommendations of *Rethinking Construction*.

Sir John Egan's report was commissioned by the Department of the Environment, Transport and the Regions (DETR) as a result of the growing dissatisfaction of both public and private clients with the performance of the industry, and the task force was asked to investigate and identify methods of improving the efficiency of the industry and the quality of its products.

The report was very critical of the industry for its poor past performance and called for a radical change in the way it implemented projects. The integration of the often separated processes of design and construction, increased standardisation, the use of lean construction, the cessation of competitive tendering and the eventual abolition of formal building contracts were seen as the main means of achieving the necessary change.

Year-on-year targets of a 10 per cent reduction in construction costs, a similar percentage improvement in productivity and the same percentage increase in turnover and profit were proposed. A 20 per cent increase in the number of projects completed on time and within budget and the same percentage cut in defects and accidents were also called for.

Criticism of the report centred around its provocative and unnecessarily hostile approach and its failure to address the needs of occasional/one-off clients and the implementers of the small- to medium-sized projects which make up a large proportion of the industry's annual workload.

In terms of the procurement process, the report's proposals advocated direct, trusting relationships being formed between clients, as the leaders of the project, and the other members of the team. It was implied that the need for more integration of design and construction might well lead to a breakdown in the existing barriers between the various parties involved in the project.

The increased use of partnering, alliances and similar co-operative procurement methods favoured by Latham appeared to be supported, although some commentators believed that the application of lean production theory, based on experience gained in the car-manufacturing industry, was central to the successful achievement of the task force's objective of increasing the construction industry's efficiency.

With government supporting and implementing its recommendations and participating in the £500 million programme of demonstration projects that were started in 1999 in order to put the proposals to a practical test, it is likely that the report's findings could be more effectively implemented than some of its predecessors.

Apart from these two major publications, a number of reports were published during the decade by government, the majority of which dealt with procurement in construction either in a general or specific way. The first of these, *Partnering: Contracting Without Conflict* [21], was published in 1991 by the National Economic Development Committee and reported on a study of the practice of partnering based upon the experience of the construction industry in the USA.

The study concluded that the technique was beneficial to clients and to the industry as a whole and that there was ample scope for the continued development of the embryo partnering activity currently taking place in the UK. A number of recommendations were made for the successful implementation of the method and a model form of contract was produced.

The following year, HM Treasury's Central Unit on Purchasing produced guidance on the selection of contract strategies (procurement systems) for major projects [22] for the use of government departments. The document, after reviewing and defining a contract strategy, described the way a project should be analysed, then explained and evaluated the various options for procuring the project, briefly outlined the selection process and provided a detailed plan for implementing the strategy.

While the terminology – particularly the use of the phrase 'contract strategy' – was somewhat confused, the setting out of a disciplined approach to the process of procurement system selection was to be applauded, especially when adopted by such a major client as central government.

In July 1995, the results of an efficiency scrutiny of the way in which government carried out construction procurement was published [23]. This examination had been commissioned by the Cabinet Office to establish the way in which government departments and agencies procured construction work and to recommend how it could ensure that it became a best-practice client.

This report should be seen as the vehicle for determining the most effective way of implementing the government's commitment given in the White Paper [24] published 2 months previously and as confirmation of its acceptance of the need to implement many of the recommendations of the Latham report.

The scrutiny looked at twenty major projects, examined fourteen procurement systems used by government departments and consulted 200 people, nearly 50 per cent of whom were in the private sector, and concluded that the UK construction industry was 'in poor shape'. Specifically, the investigating team found the industry very adversarial, without customer focus, claims conscious, internally fragmented and divided, undercapitalised with low profit margins and reluctant to adopt modern techniques for improving quality of service or the use of technology.

Government, as a major client, did not escape criticism, with all the projects that had been examined showing substantial increases over their approved costs and many failing to meet the planned completion date on time. There was a lack of understanding of the industry and insufficient involvement with the project or the consultants and contractors.

In an effort to rectify this state of affairs and to ensure that government became a best-practice client, twenty-two recommendations were made which laid down procedures for reorganising department procurement activities and methods, implementing intelligent risk management on all projects and increasing co-operation with the supply side of the industry to improve productivity and efficiency. All recommendations were accompanied by a deadline for their implementation.

Although insufficient time has passed to determine the efficacy of all, or any, of these reports, it will be seen that during the period being examined real attempts have been made by government and the industry to tackle any shortcomings and that, contrary to the lack of action which has been the fate of similar reports in past decades, efforts have and are being made to make progress on the numerous proposals for improvement – only time will tell whether, this time, words will be matched by deeds.

In this fourth and final ongoing phase, the use of design and build increased much more rapidly than during the previous two decades. Management-orientated methods fluctuated in their level of use, but generally increased their share of the market, with management contracting regaining ground so that, by value, it was back to its 1980s levels by the end of the century. The use of construction management also appears to be increasing compared with the immediate past. Partnerships and alliances appear to be on the increase, although definitive information is scarce owing to the comparative newness of the techniques used in the UK.

3.4 The level of use of procurement systems

None of the many reports produced since the 1960s on the construction industry and that have been sponsored by central government, clients or the industry itself has accurately or adequately defined the level of use by clients of the various available procurement methods that were, and are now, available.

Some of these studies did, however, give indicative figures. The Working Party, when preparing the 1967 study *Action on the Banwell Report* [8], established that in the housing and educational sections of the public sector the methods of appointing contractors were as shown in Table 3.1. It will be noted that with the exception of the 'package deal' no specific system of procurement is mentioned, although it is suggested that it would be fairly safe to assume that, with the one exception, the conventional method or its variants were used on all of the projects examined.

The level of use in the other sections of construction activity was even less definitive – hospital boards used selective tendering, universities used

Table 3.1 Method of appointing the contractor

Sector	Method of appointment	% By number of schemes	
		1964	1965
Housing	Open competition	49.2	43.2
	Selective competition	20.0	20.4
	Negotiated	23.3	27.8
	Package deal	7.5	8.6
Education	Open competition	22.4	26.2
	Invited list	21.2	21.1
	Select list	42.4	37.5
	Negotiated	14.0	15.2

selective tendering and negotiation, and the majority of contracts placed by central government and its nationalised industries were procured by selective tendering and various forms of negotiation.

The 1974 NEDO guide *Before you Build – What a Client Needs to Know about the Construction Industry* [25] confirmed that over 70 per cent of projects were still procured on a conventional basis, 18 per cent were implemented using the design and build system and the remainder were carried out by 'management' and other methods.

The Wood Report [10], which examined the purchasing policies and practices of the public client, established that 60 per cent of contracts were let on a conventional basis, nearly 25 per cent used the design and build method and the remainder were implemented using other non-conventional procurement systems.

Hillebrandt [26] assessed the level of use of procurement systems in both the public and private sectors, as shown in Table 3.2, although subsequent reports indicate that the amount of selective tendering used in the private sector was underestimated and the use of management-orientated methods was substantially overestimated.

A survey of twenty-one public- and private-sector organisations carried out in 1985 by Hewitt [27] established that the use of conventional procurement systems, mainly in the form of selective competition and negotiation, was predominant, with design and build being the next most popular method (see Table 3.3).

The *Faster Building for Industry* report [16] established that of a survey of 5,000 industrial projects undertaken during 1980–1981 just over half of these were carried out using conventional methods of procurement, one-third were implemented using design and build and the remainder by some form of management-orientated method.

The NEDO's subsequent report on the commercial sector [17] was based upon a representative sample of sixty projects, built between 1984 and 1986, supplemented by additional data gleaned from 200 schemes and a statistical

Table 3.2 Assessment of the level of use of procurement systems in the early 1980s, after Hillebrandt [26]

Procurement method	% of value	
	Private client	Public client*
Open tendering	–	10
Selective tendering	30	70
Negotiation	10	15
Design and build Two-stage tendering	} 30	} 5
Project and construction management Management contracting	} 30	

* excluding housing

Table 3.3 Most commonly used procurement systems, mid-1980s, after Hewitt [27]

Procurement strategy used	%
Conventional or variants	
selective competition	35.60
negotiation	23.74
two-stage tendering	8.47
prime cost and fee	8.47
Design and build	15.25
Turnkey	3.39
Management contracting	5.08

analysis of non-detailed information on some 8,000 commercial projects. The report concluded that over two-thirds of the contracts were let on a conventional basis, about one-sixth used the design and build approach and the remainder were carried out using some type of management procurement method.

Apart from the biennial survey carried out by the Royal Institution of Chartered Surveyors (RICS), about which more later, the lack of comparable figures for the level of use of the various procurement methods over a set period of time from a sufficiently wide range of respondents meant that it was not possible in the late 1980s to quantify accurately the actual past use of each individual system, or identify the long-term trends, on the basis of the preceding data.

In an effort to establish a more accurate picture, the author undertook a survey [28] of clients, architects and quantity surveyors. Respondents were asked to identify, by percentage of the total value of executed projects, the procurement methods they had used in the latest year for which such information was available and for the 12-month period 5 years before this. Table 3.4 shows the results of the survey, which related to industrial and commercial projects carried out during 1986 and/or 1987 and 5 years previously.

Although it can be seen that the results obtained from the three groups of respondents varied, in some cases quite radically, the trends in use for each of the procurement systems were, in the main, consistent.

The use of *conventional systems* by all of the respondents had decreased over the 5-year period in question by between 7 per cent and 14 per cent, thus confirming the trend identified by previous surveys.

The *design and build* method had gained in popularity over the surveyed period, once again confirming past evidence, with its increase in use reported as nearly 35 per cent by quantity surveyors, as over 36 per cent by clients and as 60 per cent by architects.

Taken together, *management contracting* and *construction management* showed a decrease in their use by some 34 per cent in the case of clients, an increase in their use of 54 per cent by architects and an increase of 75 per cent by quantity surveyors. This wide variation in use was not unexpected bearing in mind that at the time of the survey these two systems were mainly used on large-value, complex projects and were thus only employed intermittently by even major clients.

Design and manage had been used by clients for some 9 per cent, by value, of the construction work that they had carried out during the latest year for which they had records, and the level of use had not altered substantially from that of 5 years earlier; architects and quantity surveyors reported an increase in use during the latest year from a comparatively low base. The relatively high level of use among clients of this procurement

Table 3.4 Incidence of use, by percentage of annual value of work commissioned, of the main procurement systems

Procurement system	Clients		Architects		Quantity surveyors	
	% by value		% by value		% by value	
	latest year	5 yrs prior	latest year	5 yrs prior	latest year	5 yrs prior
Conventional	56.36	60.85	72.32	81.42	65.70	76.52
Design and build	23.64	17.36	12.55	7.84	11.13	8.26
Package deals	1.54	1.15	2.90	1.90	2.91	2.61
Turnkey	2.95	4.49	0.10	0.16	0.87	1.31
Management contracting	2.05	3.85	4.55	3.48	13.17	7.91
Construction management	1.79	1.97	4.26	2.23	1.78	0.65
BPF system	2.31	0.64	–	–	0.96	0.65
Design and manage	9.36	9.69	3.32	2.65	2.83	0.70
Other	–	–	–	0.32	0.65	1.39

method stemmed mainly from the fact that most of the respondents utilised their in-house project management expertise to co-ordinate their own internal designers, external designers and package/works contractors in order to carry out a high proportion of their construction projects on a direct basis.

The level of use of the remaining methods was so low as to preclude any meaningful conclusions being drawn from the results, although the responses obtained with regard to the use of the *British Property Federation system* appeared to confirm the opinions expressed at the time of the survey, i.e. that the take-up of this method had been slow and partial.

The RICS has, since 1984, carried out a biennial survey of contracts in use in an attempt to determine trends in the use of forms of contract and procurement methods. The survey questionnaire is sent to a random sample of private quantity-surveying practices and public bodies employing chartered quantity surveyors.

The survey attempts to sample all new-build and refurbishment work carried out in the UK in a calendar year but excludes civil engineering and heavy engineering projects, routine maintenance and repair work and work carried out by subcontractors. In 1998, the survey captured over one-fifth by value of all new UK construction orders.

The latest survey report [29] was published in January 2000, and Table 3.5 shows the trends in the use of methods of procurement identified over the 14 years since the surveys were begun.

Looking at the main categories of procurement systems, it will be seen that the use of the *conventional method and its variants* has decreased fairly evenly over the period with an overall reduction in use between 1984 and 1998 of just over 51 per cent.

The decline in use of these methods is the result of clients wanting to simplify their contact with the project team by using the resources of one organisation to carry out the design and construction of their projects, thus avoiding the need to deal with a large number of consultants and contractors and saving money in the process. Even on those projects that used the conventional approach, a significant increase in the use of contractors to design parts of the project was recorded.

Table 3.5 Trends in the methods of procurement, 1984–1998

Procurement method	Percentage							
	1984	1985	1987	1989	1991	1993	1995	1998
Conventional and variants	78.4	74.9	73.2	66.1	57.8	54.0	58.3	40.2
Design and build	5.1	8.0	12.2	10.9	14.8	35.7	30.1	41.4
Management contract and fixed fee	16.5	17.1	14.6	16.1	8.0	6.4	7.4	10.7
Construction management	–	–	–	6.9	19.4	3.9	4.2	7.7
Total	100	100	100	100	100	100	100	100

Design and build itself, after a steep but steady climb during the 1980s and early 1990s, showed a dramatic increase in use during the remainder of the last decade and an overall growth of over 800 per cent during the time of the surveys.

The use of *management contracting and prime cost plus fixed fee* arrangements remained fairly constant between 1984 and 1989, declined by half in 1991 and has stayed at about the same level since that time. The sharp fall in use at the end of the 1980s probably reflected the disillusionment of clients with the method as well as the start of the recession that overtook the industry at that time.

It is difficult to judge the changes in use of *construction management* by comparing biennial value as the financial size of many projects tends to distort the figures. Looking at the *number* of projects carried out by this method over the 9 years since 1989 there has been a reasonably steady rise in use since that time, with four times as many projects being carried out in 1998 as in the first year for which figures were collected.

The survey also comments on two other forms of procurement: *partnership or alliance* arrangements, in which nearly 2 per cent of the 2,457 projects captured had made such provision, and a *private finance/public private partnership* approach, which was used in schemes with a combined value of £215 million, i.e. 4.5 per cent of the value of the total projects surveyed. The latter figure must be conditioned by the fact that in 1998 very few of this type of arrangement had been started on site and a considerable growth in value and numbers should be expected in the next survey.

It is not possible, on the basis of literature and past research, to quantify accurately the past or present level of use of all, or any, of the available procurement systems. This inability stems from the lack of truly comparative figures for the individual methods over a set period of time from a sufficiently wide range of reliable sources.

The RICS surveys come nearest to achieving these criteria and, together with those other surveys that most nearly meet this standard, form the basis of Table 3.6, which attempts to provide an indication of the pattern of use of such systems.

Figure 3.2 shows the difference in the level of use of three categories of procurement system over a period of nearly 40 years between 1960 and 1998, as estimated by Rowlinson [30] and the latest RICS survey [29]. Despite the industry's continued reluctance to accept and implement fully the recommendations of numerous reports made over nearly 40 years, the figure illustrates the changes that have occurred mainly as a result of the determination of an increasing number of clients to implement their projects by the use of procurement methods which satisfy their own needs rather than those of the construction industry.

The next four chapters define and describe the various procurement systems and their variants and provide information on the history and characteristics of the methods, as well as the process of implementation and the advantages and disadvantages of the different systems.

Table 3.6 Pattern of use of procurement systems, 1981–1998

Year of actual study	Design and build and variants (%)	Conventional and variants (%)	Management and other (%)	Reference
1981	52	34	14	16
1982–1983	61	23	16	28
1984–1986	67	16	17	17
1984	78	5	17	31 (1985)
1985	75	8	17	31 (1986)
1986	73	12	15	31 (1988)
1987–1988	56	28	16	28
1989	66	11	23	31 (1990)
1991	58	15	27	31 (1992)
1993	54	36	10	31 (1994)
1995	58	30	12	31 (1996)
1998	40	41	19	31 (2000)

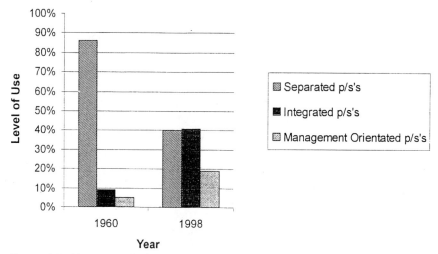

Figure 3.2 Changes in the level of use of procurement systems, 1960–1998

References

1 McDermott, P. (1999) 'Strategic and emergent issues in construction procurement', in *Procurement Systems: A Guide to Best Practice in Construction*, Rowlinson, S.M. and McDermott, P. (eds), London: E & FN Spon.

2 Clamp, H. and Cox, C. (1990) *Which Contract. Choosing the Appropriate Building Contract*, London: Royal Institute of British Architects Publications.

3 Perry, J.G. (1985) 'The development of contract strategies for construction projects', unpublished PhD thesis, University of Manchester.

4 Central Council for Building and Works (1944) *The Placing and Management of Building Contracts (The Simon Report)*, London: HMSO.

5 Ministry of Works (1950) *Report on the Working Party on the Building Industry (The Phillips Report)*, London: HMSO.

6 Ministry of Works (1962) *Survey of Problems Before the Construction Industry (The Emmerson Report)*, London: HMSO.

7 Ministry of Public Building and Works (1964) *The Placing and Management of Contracts for Building and Civil Engineering Work (The Banwell Report)*, London: HMSO.

8 Economic Development Committee for Building (1967) *Action on the Banwell Report*, London: National Economic Development Office.

9 Higgins, G. and Jessop, N. (1965) *Communications in the Building Industry*, London: Tavistock Publications.

10 Building Economic Development Committee (1975) *The Public Client and the Construction Industry (The Wood Report)*, London: HMSO.

11 Slough Estates (1976) *Industrial Investment. A Case Study in Factory Building*, London: Slough Estates.

12 Slough Estates (1979) *Industrial Investment. A Case Study in Factory Building*, London: Slough Estates.

13 Building and Civil Engineering Development Committees (1978) *Construction for Industrial Recovery*, London: HMSO.

14 University of Reading, Department of Construction Management (1979) *UK and US Construction Industries: a Comparison of Design and Contract Procedures*, London: Royal Institution of Chartered Surveyors.

15 British Property Federation (1983) *Manual of the BPF System for Building Design and Construction*, London: British Property Federation.

16 Building Economic Development Committee (1983) *Faster Building for Industry*, London: National Economic Development Office.

17 Building Economic Development Committee (1988) *Faster Building for Commerce*, London: National Economic Development Office.

18 Cox, A. and Thompson, I. (1998) *Contracting for Business Success*, London: Thomas Telford.

19 Latham, M. (1994) *Constructing the Team: Joint Review of Procurement and Contractual Arrangements in the United Kingdom Construction Industry*, London: HMSO.

20 Department of the Environment, Transport and Regions (1998) *Rethinking Construction. The Report of the Construction Task Force (The Egan Report)*, London: HMSO.

21 National Economic Development Committee (1991) *Partnering: Contracting Without Conflict*, London: HMSO.

22 HM Treasury, Central Unit on Purchasing (1992) *No. 36 Contract Strategy: Selection for Major Projects*, London: HMSO.

23 Cabinet Office Efficiency Unit (1995) *Construction Procurement by Government*, London: HMSO.

24 HM Treasury (1995) *Setting New Standards – A Strategy for Government Procurement*, London: HMSO.

25 Building Economic Development Committee (1974) *Before you Build – What a Client Needs to Know About the Construction Industry*, London: National Economic Development Office.

26 Hillebrandt, P.M. (1985) *Economic Theory and the Construction Industry,* 2nd edn, Basingstoke: Macmillan.
27 Hewitt, R.A. (1985) 'The procurement of buildings: proposals to improve the performance of the industry', unpublished project report submitted to the College of Estate Management for the RICS Diploma in Project Management.
28 Masterman, J.W.E. (1989) 'The procurement systems used for the implementation of commercial and industrial building projects', unpublished MSc thesis, University of Manchester Institute of Science and Technology.
29 Davis, Langdon and Everest (2000) *Contracts in Use: A Survey of Contracts in Use During 1998,* London: Royal Institution of Chartered Surveyors.
30 Rowlinson, S. (1986) 'An analysis of the performance of design–build contracting in comparison with the traditional approach', unpublished PhD thesis, London: Brunel University.
31 Royal Institution of Chartered Surveyors (1985, 1986, 1988, 1990, 1992, 1994, 1996, 2000) *Contracts in Use in 1984, 1985, 1987, 1989, 1991, 1993, 1995, 1998,* London: Royal Institution of Chartered Surveyors.

4 Separated procurement systems

4.1 Introduction

The unique characteristic of this category is the separation of the responsibility for the design of the project from that of its construction. The category contains one procurement system – the *conventional method*.

4.2 Definition of the system

This method of procuring building projects is usually referred to within the industry and literature as 'the traditional method'. The author maintains that, in fact, the traditional method of designing and constructing buildings was that which was most commonly used before the end of the 1700s, or beginning of the 1800s, when clients had for many centuries traditionally employed craftsmen, on an individual basis, under the supervision of a master mason or surveyor or, very rarely, an architect.

Since that time and until relatively recently, the 'accepted social behaviour' – the definition of 'convention' in *The Concise Oxford Dictionary* – has been for most clients to implement their building projects by using a main contractor, with the design and supervision being carried out by an architect assisted by other specialist consultants. The term 'conventional method' has, therefore, been used to describe this system throughout this guide.

Apart from the separation of design and construction, the conventional procurement system exhibits a number of other basic characteristics:

1   Project delivery is a sequential process.
2   The design of the project is largely completed before work commences on site.
3   The responsibility for managing the project is divided between the client's consultants and the contractor, and there is therefore little scope for involvement of either of the parties in the other's activities.
4   Reimbursement of the client's consultants is normally on a fee and expenses basis, whereas the contractor is paid for the work completed on an admeasure or lump sum basis.

The ideal definition of this method will include all of these features, and the following attempt at encapsulation has been adopted for the purpose of use within this guide.

> The client appoints independent consultants, on a fee basis, who fully design the project and prepare tender documents upon which competitive bids, often on a lump sum basis, are obtained from main contractors. The successful tenderer enters into a direct contract with the client and carries out the work under the supervision of the original design consultants.

Figure 4.1 illustrates the conventional procurement system in simplified diagrammatic form.

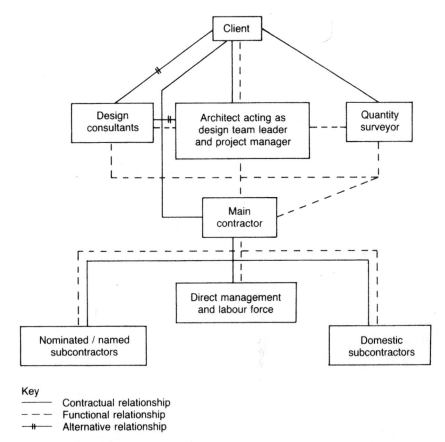

Key
——— Contractual relationship
– – – Functional relationship
—H— Alternative relationship

Figure 4.1 Contractual and functional relationships, the conventional system of procurement

4.3 Genesis

The conventional procurement system has been used by the majority of clients of the industry for at least the past 150 years in order to implement their building projects.

Powell [1] maintains that a watershed in the use of the system occurred during the Napoleonic wars, (1792–1815), when, as a result of the frequent disputes that arose between the client and the separate tradesmen involved in the 'measure and value' system current at that time, the government's Office of Works introduced competitive tendering for entire projects as a superior alternative.

By the 1850s, 'contracting in gross', as the system was then known, widely prevailed, enabling clients to secure the economic benefits of competition, knowledge of the final cost before work began, better control of subsequent expenditure and the ability to enter into a single, contractual relationship with a builder instead of tradesmen, who were less co-ordinated. At the same time as the organisations for implementing projects and the associated contractual relationships were changing, surveyors were being used by groups of builders, and eventually by clients, to take off the quantities of materials required for estimating purposes from the architect's drawings, thus relieving the tenderers of responsibility for their accuracy and sufficiency.

From the second half of the nineteenth century until the commencement of the Second World War, the use of the conventional system by both public and private clients gradually increased to the point where, in England and Wales at least, this method was being used for the vast majority of projects, albeit on the basis of very open competitive tendering.

After the end of hostilities in 1945, the conventional procurement system remained predominant, supported by the findings of the Simon committee [2], which had reported that an examination of the system of placing contracts had not disclosed any serious weakness in the methods that had been built up by 'architects, quantity surveyors and builders', although it was recommended that, where appropriate, selected and limited lists of tenderers should be used together with negotiation.

The story of that most demanding of clients, the commercial property developer, from 1945 to 1967 as told by Marriott [3] makes no mention of the use of any procurement system other than the conventional method, apparently confirming that, until at least the mid-1960s, this approach satisfied the majority of the metropolitan property companies.

The seeds of change, however, had already been sown in the Emmerson and Banwell Reports of 1962 and 1964 [4,5]. Emmerson concluded that ways needed to be found to improve co-ordination and co-operation among the building owner, consultants, contractors and subcontractors, and suggested that the system for placing contracts and managing projects should be comprehensively reviewed.

Banwell [5] reiterated this view and berated the majority of the various members of the industry for their reactionary approach to new ideas and processes, pointing out the urgent need for the separate factions to come together and think and act as a whole, particularly in the 'letting, form and management of contracts'.

Perhaps as a result of the pressure exerted by such reports, as well as by normal commercial influences, the mid- to late 1960s were a watershed in terms of the increased use of non-conventional procurement systems and thus the decline in the employment of conventional methods.

This period saw the beginning of the growth of design and build, the birth of the first major management contract, the use of serial tendering for system build/industrialised buildings and a general acceptance among the larger and more far-sighted clients and consultants that the involvement of the contractor at an early stage could be of benefit to the project as a whole.

In 1973, the oil crisis, and the consequent rise in fuel prices and interest rates, meant that the dominant project objective of most clients became the need for rapid commencement and completion of a development in order to reduce borrowing to a minimum. These requirements resulted again in a continuing decline in the use of conventional procurement systems.

This trend has continued until the present time, with clients satisfying their needs by increasingly using non-conventional procurement systems to the detriment of the conventional method.

4.4 Share of the market

While a great deal of time and effort has been expended in achieving technical innovation in construction, and refining existing and producing new forms of contract, very little attention had been paid, until comparatively recently, to the rationalisation and reorganisation of the procurement process, thus allowing the conventional system to maintain a major share of the available building work.

However, the increasing participation of large and sophisticated clients in this process, particularly during the past 20 or so years, and the historical events that have previously been described has meant that the amount of work being carried out using the conventional procurement method has declined, a trend that appears likely to continue.

The lack of accurate comparative historical data on the level of use of procurement systems and the wide range of types of projects and categories makes any examination of this aspect of construction activity extremely difficult and any subsequent results somewhat unreliable.

The wide variation in the estimates of the level of use of this system means that any attempt to evaluate the share of the market currently enjoyed by the conventional method, and its offshoots, is fraught with difficulty, but it is known that the level of use had shrunk from an estimated 85 per cent in 1960 to something around 40 per cent at the end of the 1990s.

At the present time, the market's perception of the conventional system and its variants is that for small- and medium-sized uncomplicated projects where time is not at a premium and where the client wishes to have continuous control over the design this system offers the most economic route to success. In the case of major new works, where time is at a premium, the shortcomings of the separated methods are likely to result in the use of non-conventional systems.

In the immediate future, and as long as the present high level of activity continues, it is likely that this method will continue to lose ground, although perhaps at a slower rate than during the past three decades. However, in the longer term, it could be argued that, if demand reduces and the past cyclical pattern of building industry activity is repeated, the rate of decline in the use of conventional systems may reduce and even stabilise.

4.5 The process

The outline plan of work drawn up by the Royal Institution of British Architects (RIBA) [6] sets out the procedures to be followed for projects being procured by conventional means. Although the plan identifies twelve stages, there are only four which need concern us in the context of procurement systems. These are *preparation, design, tender* and *construction*.

As one of the unique characteristics of the conventional system is that it follows a strictly sequential path, each of these four stages can be viewed as separate entities and carried out, to a certain extent, in isolation of the others, with the result that the process can become extremely lengthy, lead to poor communication, undermine relationships between project team members and produce problems of buildability. The four stages are now described and discussed.

Preparation

This is the inception stage of the project when the client establishes his/her needs in principle, but not in detail, appoints a project manager and selects and appoints a design team, which normally consists of, as a minimum, an architect, a structural/civil engineer, a mechanical and an electrical engineer and a quantity surveyor, together with any other specialist consultants necessary for the successful implementation of the project. Dependent upon the nature of the client, and the undertaking itself, the project manager may be:

- an employee of the client organisation with no construction knowledge who simply acts as a co-ordinator of information and a single source of contact and communication for the design team leader, who will have responsibility for the day-to-day project management;

- an experienced construction professional, permanently and directly employed by the client, who, in addition to being the single point of contact, will be responsible for the financial, technical and administrative management of the project from inception to completion;
- an external consultant project manager appointed, probably on a percentage fee/target-cost basis, for a specific project to carry out the same duties as an 'in-house' project manager.

It can be seen that the conventional procurement system involves the client in a number of differing relationships with several organisations, and many inexperienced customers are dismayed at the complexity of the process and at the size and cost of employing the design team itself.

During the preparation stage and before the appointment of the design team, but not the project manager, the client establishes his/her basic needs in terms of the functionality and the quality of the project and the cost and time parameters that he/she wishes to set. Having settled these fundamental requirements, the strategies that will be used to implement the project successfully will be determined and, on the basis of these, an appropriate design team will be appointed.

Any time taken during this stage to ensure that the client's requirements are correctly established will be time well spent and should be reflected in the ability to proceed with the other three phases of the project with the minimum of change and disruption. The decisions taken at this time set the whole tone and pattern for the remainder of the building process.

At this stage, and to a slightly lesser extent during the design phase of the project, the client has a great deal of influence – he/she has much less opportunity to control any aspects of the undertaking during the last two stages and particularly during the construction period.

Design

This phase sees the appointment of the design team, which develops the project through a series of stages: briefing, feasibility, outline design, scheme design and detailed design with the scheme's configuration and features becoming firmer at each stage. Again, the client and his/her consultants have considerable freedom during this phase to conceive and develop the project without excessive time or economic pressures, although research [7] has shown that projects which move quickly in the preconstruction period tend to be constructed quickly as well.

When using the conventional procurement system, it has been established that very often work has been put in hand without the client having comprehensively established his/her needs, and the design team having clarified them, with the result that expensive and disrupting changes have been necessary during the construction period.

This less-than-rare phenomenon stresses the need to ensure that the proposed design satisfies the functional and qualitative needs of the client before obtaining tenders and commencing work on site – the lack of attention to this requirement will result in an unsuccessful project.

During the design process, the designers are usually working in isolation, far removed from the contractor who will eventually be responsible for carrying out the construction of the project and sometimes from each other. This isolation from the contractor is deliberate and great care is taken to ensure that no contact occurs, thus reinforcing the division identified by numerous reports over the past four decades.

As a consequence, opportunities for incorporating and ensuring buildability, as recommended by all authorities, are virtually non-existent, although it has been suggested, perhaps somewhat naively, that even if a contractor is chosen after the design has been formulated he/she should be invited, when appointed, to suggest any modifications which might improve buildability, speed of construction or cost.

The main reasons for the lack of involvement of contractors when using the conventional method are well known but nevertheless deserve reiteration:

1   Clients wish to ensure that the responsibility for the design of any project is vested in one group, i.e. the design consultants.
2   The list of tenderers will generally not be available until the design has been mainly completed.
3   The practical and ethical difficulties of dealing with suggestions from a number of contractors during the design stage are difficult to overcome.
4   Once design decisions have been made, they usually cannot be changed without cost penalties and delays to the construction programme.

In this context, Perry [8] has identified the impact by reference to an unidentified Swedish study of work carried out by a client and his/her consultants up to and including the design stage as influencing 90 per cent of the construction cost with only 15 per cent of the actual project expenditure having been incurred. It can thus be seen that, although the client may be anxious to see work commence on site, progress during this stage should be carefully controlled and not unreasonably forced. Hastily prepared design details can lead to major misunderstandings and disputes during the construction stage, which may result in delays and cost penalties.

Preparing and obtaining tenders

Tender documentation on conventionally procured projects normally consists of drawings, specification(s) and a bill of quantities, with the last document being prepared by the quantity surveyor on the basis of measurements 'taken off' the designers' drawings in order to provide each tenderer with a common base from which to price his/her bid. For the conventional system to operate successfully, and to minimise the financial risk to the client, it is imperative that the design is fully developed before the bills of quantities are prepared and tenders invited. If this is not done, excessive variations and disruptions of works are likely to occur.

Although selection of the contractor by limited competitive tendering should offer the assurance of achieving the lowest price for the project, in reality the designers' drawings are rarely in sufficient detail to enable a bill of quantities to be prepared with any accuracy, and the art of evaluating from the drawings the exact amount of work required ranges from the difficult to the impossible.

It has been repeatedly demonstrated that, where tender documentation is suspect in this way, bids obtained by the conventional method can only be considered as indicative of the final cost to the client and may result in an unscrupulous contractor abusing the system by submitting an unrealistically low bid and then formulating claims for additional reimbursement in order to uplift his/her original tender to a commercially viable level.

The selection of contractors who will receive tender documents and submit bids can be made in a number of ways, but it is usual in the case of conventionally managed projects – particularly if the expenditure of public money is involved – for tenderers to be selected from a list of approved contractors, with only the occasional project and those governed by European Union regulations being the subject of an advertisement inviting contractors to bid.

It has now been generally accepted by all categories of client that, when using the conventional procurement system, the selection of the most acceptable bid should not be made by the use of open competitive tendering but by selective tendering based upon a list of tried and tested contractors whose performance and financial stability are regularly monitored.

A critical element of the tender documentation is the form of contract – a subject about which much has already been written; suffice it to say that in the context of procurement systems and the achievement of project success it is normal and advisable to use the most appropriate of the many standard forms of contract of which the industry has much experience.

If for any particular reason it is intended to introduce any special conditions of contract, to use non-standard agreements proposed by one of the parties or, more rarely, to formulate a bespoke form of contract, technical and legal advice is taken and the advantages and disadvantages of the proposed documentation carefully considered before any commitment is made.

Once tenders are received, the selection of the bid of most advantage to the client presents very little difficulty when using this system as tenders will be judged on price alone, having been based on documentation which is common to all tenderers and which, theoretically at least, accurately and comprehensively reflects the client's actual requirements.

Construction

When using the conventional system of procurement, an adequate period is needed for the contractor to plan the project thoroughly and organise the required resources. Undue haste in making a physical start on site may result in managerial and technical errors being made by both the design team and the contractor, which could lead to a lengthening rather than a reduction of the construction period.

It has already been established that, when using this method, a very high proportion of the estimated cost of the project has been committed before work commences on site, although actual expenditure is comparatively small. However, it is during the construction phase that the majority of difficulties will surface, with the quality of the performance during this period having already been largely determined by the quality of the preparatory work. It is at this stage that the price for an incomplete design, inaccurate bill of quantities, poorly prepared tender documentation and lack of 'buildability', etc. is paid.

The ability to introduce changes to the design of the project during the construction period is a characteristic of this system that is both a strength and a weakness as such variations have been identified as one of the most important causes of, and excuses for, delay. If it is essential to instigate changes, the project team (including the contractor) needs to be consulted and the practical and financial consequences of the proposed variation established in detail before instructions to proceed are given.

Developers building on a speculative basis have been identified as allowing their projects to be disrupted substantially by their efforts to respond to demands from prospective tenants during the construction period. The in-built ability of the system to accommodate variations can lead to a permissive attitude to design changes, and any alterations to the original design need to be kept to a minimum or, if possible, entirely eliminated.

The management and supervision of the work on site to ensure that it conforms to the client's brief as reflected in the design, specification and contract conditions is the responsibility of the design team, although it should be remembered that under normal terms of engagement the team is not required to carry out full-time supervision of the works. This is usually an additional service provided by a resident architect, engineer and/or clerk of works, who is employed at the client's expense.

The ability of design consultants in general, and architects in particular, to manage projects has been continually questioned over the past three decades, when it has been maintained that in the case of the conventional approach designers are not motivated to give sufficient attention to the control of the critical criteria of cost and time and have not been trained to manage such projects effectively.

The combination of part-time supervision and lack of management expertise and motivation during the construction phase of conventionally procured projects can result in delays and additional costs being incurred by the client as a consequence of poor performance by his/her consultants. The detailed and continuing involvement of the client can offset these deficiencies as it has been amply demonstrated that customers who take a constructive and objective interest in all aspects of their projects achieved the best results, particularly in terms of speed of completion.

Because of the separated nature of the conventional method of procurement, it is necessary for the client to ensure that good communications exist between all members of the project team, that immediate decisions are made when queries arise during the construction phase and that a strong site-management team is in place before work commences on the project.

Payment to the contractor for work that has been satisfactorily completed is made by means of interim certificates – generally monthly – to the value of work done, issued by the architect on the recommendation of the quantity surveyor. The priced bill of quantities submitted by the contractor at tender stage forms the basis of interim valuations and also ensures that any variations can be valued by reference to preagreed rates for appropriate operations. An agreed percentage is retained until final completion and a further reduced amount until the defects liability period is satisfactorily completed.

This rigid monthly payment system, together with the practice of holding retention monies, has been criticised as being a very expensive method of payment for the large number of undercapitalised small subcontractors who now carry out the majority of the actual site work, and it has been suggested that the adoption of more flexible methods of payment and the streamlining of the retention system would benefit the client as well as the industry.

The likelihood of such a major change in payment methods being achieved on all projects within the foreseeable future is remote, but in the meantime many clients are ensuring that certified payments are promptly honoured and that the subsequent reimbursement of all subcontractors by the main contractor is made with the same alacrity. Such action will ensure that the contractor's efforts are more effectively engaged in actually managing the project, rather than pursuing outstanding payments, and should serve to ensure good relations and thus improve project efficiency.

The nature of the conventional method of procurement and its associated conditions of contract is such that any delays that occur during the construction phase that are caused by events outside the main contractor's control can only be overcome or mitigated by issuing instructions to accelerate the appropriate current or future critical operations at the client's expense.

The sequential characteristic of the system reduces the ability to deal with any unexpected delays other than during the construction period. Overcoming such delays during this phase of the project is not easily achieved, even if the cost of the necessary acceleration can be accommodated within the financial budget for the scheme, and the project team needs to monitor closely the contractor's progress so that any areas of possible delay can be detected sufficiently early to enable remedial action to be taken and practical completion achieved in accordance with the client's requirements.

4.6 The product

Cost

It is maintained that when using conventional procedures, where bills of quantities form part of the tender documentation, the cost of tendering is reduced, the quantitative risks encountered in tendering are removed, competition is ensured, post-contract changes can be implemented at a fair and reasonable cost and clients can be confident that they know their financial commitment. All this is true, provided that the design has been fully developed and accurately billed before obtaining tenders. If, however, these criteria have not been strictly met, excessive variations, disruption of the works and a consequent increase in the tendered cost will occur.

There appears to be a lack of up-to-date accurate information on the actual final costs of projects carried out using the conventional method, although there have been a number of reports and guides published since the mid-1970s which touch upon this fundamental aspect of procurement management.

The Building Economic Development Committee of NEDO produced a guide [9] in 1974 in which the performance of projects designed by different types of designers was measured. Only 26 per cent of all of the factory projects examined that had been designed solely by architects or by architects assisted by other designers – in other words, procured by conventional means – were completed within 5 per cent of the estimated cost. The remainder exceeded the budget figure by a greater margin. Office projects fared little better, with 33 per cent of the projects that used the conventional system being completed within the 5 per cent figure and the remainder again exceeding the estimated cost by more than that percentage.

In 1982, the Department of Trade and Industry and the Department of the Environment jointly produced *A Guide to Methods of Obtaining a New Industrial Building in the United Kingdom* [10], which demonstrated that the sequential traditional path, i.e. the conventional method of procurement, provides a high degree of price certainty and competition.

Flanagan [11] illustrated the apportionment of financial risk between the client and the contractor when using various procurement systems (see Figure 4.2) and showed that the basic form of reimbursement associated with the conventional procurement system, i.e. the lump sum fixed price, allocated the major part of the risk to the contractor, thus, theoretically at least, reducing the cost of the project to the client to the minimum.

Brandon *et al.* [12] assumed values for an addition to the unit cost of the project to account for the effect of the chosen procurement system on the contract sum. On a competitive lump sum contract, with a full bill of

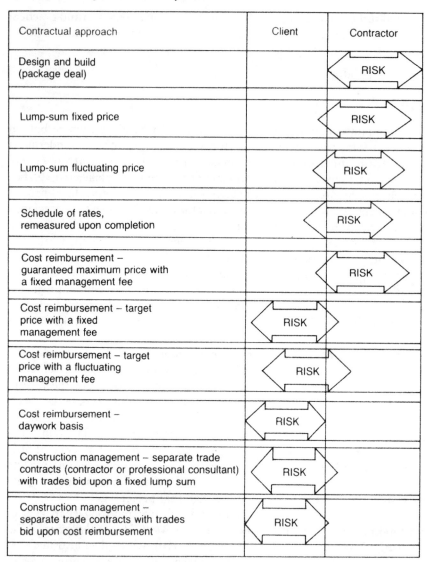

Contractual approach	Client	Contractor
Design and build (package deal)		RISK
Lump-sum fixed price		RISK
Lump-sum fluctuating price		RISK
Schedule of rates, remeasured upon completion		RISK
Cost reimbursement – guaranteed maximum price with a fixed management fee		RISK
Cost reimbursement – target price with a fixed management fee	RISK	
Cost reimbursement – target price with a fluctuating management fee	RISK	
Cost reimbursement – daywork basis	RISK	
Construction management – separate trade contracts (contractor or professional consultant) with trades bid upon a fixed lump sum	RISK	
Construction management – separate trade contracts with trades bid upon cost reimbursement	RISK	

Figure 4.2 Allocation of financial risk

quantities, the addition was valued at 0 per cent, which, while being equal to that of the design and build system, was the lowest of the remaining five procurement methods that were examined.

Turner [13] suggests that the project price may be higher as a result of the longer period of design and construction, but does not make any comparison with the cost of using other systems of procurement.

In more general terms, the project cost can be estimated, monitored and controlled by the client's cost advisor during the whole life of the project –

an important advantage in the case of large cost overruns and as a means of ensuring that they do not occur in the first place.

Such data as are available would therefore appear to support the widely held belief that the use of the conventional procurement system results in a final project cost which is lower than if any of the other methods (with the exception of design and build) had been used, subject of course to the tender documentation being based upon a fully completed design.

Time

Because of its sequential nature, the conventional procurement system has been continuously identified as the slowest method of procuring construction projects available to a client. An examination of research into this aspect of the method supports this belief far more readily than with the previous cost characteristics of the method.

The 1974 Building Economic Development Committee (BEDC) guide [9] showed that, of the sampled factory projects carried out using the conventional method, nearly 55 per cent were completed within 5 per cent of the programmed time for the design element, and 51 per cent were completed for the construction element. Similar figures for office buildings in the same study revealed that nearly 60 per cent of the projects were completed within 5 per cent of the estimated design period, and nearly 54 per cent were completed within the same percentage of the estimated construction period. Thus, between 40 per cent and 49 per cent of all the projects surveyed overran their estimated design and construction period by in excess of 5 per cent.

The 1975 Wood Report [14] also established that, of a sample of 2,000 public-sector building and civil engineering projects which were examined, over 80 per cent were implemented by the use of the conventional procurement method. Although no specific results were obtained for this major element of the sample, over 60 per cent of the total projects overran by an average of 17.4 per cent, with one-third exceeding the estimated project period by more than 20 per cent.

A report produced by NEDO in 1978 [15] surveyed over 500 industrial companies, the majority of which had carried out construction projects using the conventional procurement system and found that of the 300 or so who had recent capital investment experience 17 per cent were dissatisfied with the project times achieved by the construction industry.

Figure 4.3 illustrates the comparison between the times achieved by the various listed procurement methods during the preconstruction stage of the projects surveyed in the 1983 NEDO Report *Faster Building for Industry* [7].

In addition to the time saved by the use of variants of the conventional system during the preconstruction phase of the project, it was established that further savings could be made during the construction phase itself by

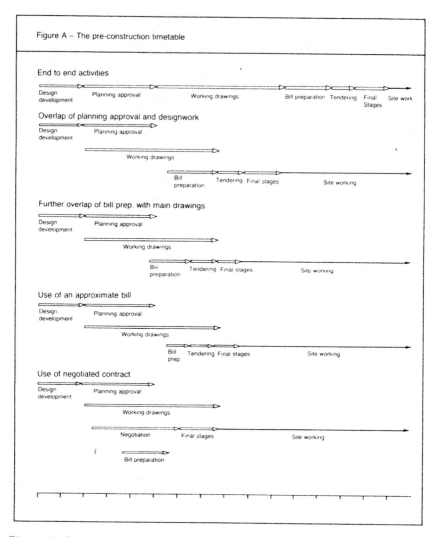

Figure 4.3 Preconstruction times

overlapping design and construction when using other non-conventional systems, and the report therefore concluded that:

> ... the use of non-traditional routes tends to produce overall project times shorter than those produced by traditional routes.

Quality and functional suitability

The generally held view among clients is that the conventional procurement system provides a high degree of certainty that quality and functional standards will be met. This view is supported by the findings of a BEDC guide [9] which examined the existence of faults in completed buildings when designed by different designers working within various procurement systems.

On factory buildings, designed by architects employed directly by clients, 49 per cent were free of faults, 40 per cent had minor defects and 4 per cent had major problems. In the case of office buildings designed under the same circumstances, 54 per cent had no faults, 41 per cent had minor defects and 5 per cent had major defects.

The 1982 guide *Methods of Obtaining a New Industrial Building in the UK* [10] identified the conventional procurement method as being suitable for projects of normal to more demanding levels of building quality.

The study *Faster Building for Industry* [7] found that clients who built speculatively used conventional procurement arrangements because they were concerned about the market appeal of their buildings and wanted to retain control over appearance, quality of finishes, etc.

Finally, the Construction Round Table's guide *Thinking About Building* [16] reinforces the suitability of the conventional route for those whose projects are complex or who require high-quality or prestigious buildings.

Other characteristics

The conventional procurement system has the advantage of having stood the test of time over many years and being understood by many clients and by all the participants from the construction industry itself. The client is able to select the most appropriate design team for his/her project, taking advantage of their experience of similar developments, and can also delay commitment to a building contract until a later stage in the development of his/her requirements.

The main disadvantages of the system have already been highlighted, but there is little doubt that the conventional procurement method does not motivate the client to make decisions as firmly, or as early, as he/she should nor to induce designers and contractors to pay enough attention to saving and controlling cost, or time, and improving building quality. It is also the case that the designers of the project often have no direct experience of managing construction work and the contractor is unable to contribute to the design of the project until too late.

However, it should be said at this point that enormous pressure is usually needed to change procedures, such as the conventional procurement method, which become institutionalised within the industry, but during the past decade such pressure has been exerted, mainly by the large property developers, to the point where some would argue that non-conventional procedures have been used inappropriately and the conventional system discarded unnecessarily.

What needs to be sought is a balance between the use of proven non-conventional methods and the known ability of the conventional system to offer the most economic procurement route for small- and medium-sized uncomplicated projects where time is not at a premium.

4.7 Summary

The separated category has one fundamental and detrimental characteristic in that the responsibility for the two main elements of 'design' and 'construction' is vested in two separate organisations – the design team and the contractor.

The advantages and disadvantages of the conventional procurement system are as follows.

Advantages

1   Provided that the design has been fully developed and uncertainties eliminated before tenders are invited, tendering costs are minimised, proper competition is ensured, the final project cost will be lower than when using the majority of other procurement methods and the selection of the bid that is most advantageous to the client will present little difficulty.
2   The existence of a priced bill of quantities enables interim valuations to be assessed easily and variations to be quickly and accurately valued by means of preagreed rates.
3   The use of this method provides a higher degree of certainty that quality and functional standards will be met than when using other systems.

Disadvantages

1   Where tenders are obtained on the basis of an incomplete design, the bids obtained can only be considered as indicative of the final cost and the client is thus vulnerable to claims for additional financial reimbursement from the contractor.
2   The sequential, fragmented and confrontational nature of this system can result in lengthy design and construction periods, poor communication between clients and the project team and problems of buildability.
3   While the facility to respond to late demands for change, by introducing variations, can result in satisfied customers, such action has been identified as one of the main causes of delay, and increased cost, and can lead to a permissive attitude to design changes.

Chapter 9 is devoted to the way in which these characteristics are taken into account when selecting the most appropriate building procurement system for specific client organisations and projects, but in general terms the use of the conventional method should be favoured when speed is not a critical factor, costs need to be minimised and quality and functionality assured.

References

1 Powell, C.G. (1980) *An Economic History of the British Building Industry, 1815–1979*, London: Methuen.
2 Central Council for Building and Works (1944) *The Placing and Management of Building Contracts (the Simon Report)*, London: HMSO.
3 Marriott, O. (1967) *The Property Boom*, London: Hamish Hamilton.
4 Ministry of Works (1962) *Survey of Problems before the Construction industries (The Emmerson Report)*, London: HMSO.
5 Ministry of Public Buildings and Works (1964) *The Placing and Management of Contracts for Building and Civil Engineering Work (the Banwell Report)*, London: HMSO.
6 Royal Institute of British Architects (1995) *Architect's Job Book*, London: Royal Institute of British Architects Publications.
7 Building Economic Development Committee (1983) *Faster Building for Industry*, London: National Economic Development Office.
8 Perry, J.G. (1987) 'Course notes', MSc in Construction Management, University of Manchester Institute of Science and Technology.
9 Building Economic Development Committee (1974) *Before you Build – What a Client Needs to Know About the Construction Industry*, London: National Economic Development Office.
10 Department of the Environment (1982) *The United Kingdom Construction industry – a Guide to Methods of Obtaining a New Industrial Building in the UK*, London: Invest in Britain Bureau.
11 Flanagan, R. (1981) 'Change the system', *Building* 20 March, 28–29.
12 Brandon, P.S., Basden, A., Hamilton, I.W. *et al.* (1988) *Expert Systems. The Strategic Planning of Construction Projects*, London: Royal Institution of Chartered Surveyors.
13 Turner, A. (1997) *Building Procurement*, 2nd edn, Basingstoke: Macmillan.
14 Building Economic Development Committee (1975) *The Public Client and the Construction industries (The Wood Report)*, London: National Economic Development Office.
15 Building and Civil Engineering Development Committee (1978) *Construction for Industrial Recovery (the Graves Report)*, London: National Economic Development Office.
16 Construction Round Table (1995) *Thinking About Building*, London: Business Round Table.

5 Integrated procurement systems

5.1 Introduction

This category of procurement systems incorporates all of those methods of managing the design and construction of a project where these two basic elements are integrated and become the responsibility of one organisation, usually a contractor. In the case of one variant, other elements, such as funding and operating, are also incorporated within the system.

The *design and build* procurement system is the main member of this category, with variants of that method making up the remainder of the group. The principal variants that are dealt with here are *novated design and build*, the *package deal*, *develop and construct*, and *turnkey* methods of procurement.

Various authorities have suggested that the design and manage and the British Property Federation (BPF) systems should be included within this category. The former method, although exhibiting many of the characteristics associated with true design and build, is fundamentally different in one particular respect in that, when taking the design and manage route, the implementing organisation usually acts in a consultant capacity and carries out the design and management of the project on a fee basis, using works/package contractors to execute the actual construction operations. The BPF system, while making provision for the contractor to undertake certain design and constructional detailing, does not necessarily place the responsibility for the design and construction of the project with one single organisation. For these reasons, these two systems are dealt with in what are considered to be their proper categories in other more appropriate chapters, and only design and build itself, and its four variants, are now discussed and examined.

5.2 Design and build

Definition

The term 'package deal' has been used in the building industry for many years as an all-embracing description covering design and build, the 'all-in' service, develop and construct, and turnkey contracting.

This imprecise terminology has led in the past to considerable confusion among practitioners and clients of the industry, resulting in misunderstandings as to the nature, characteristics, advantages and disadvantages of the basic design and build system and, in some extreme cases, to its rejection as a possible candidate for use on some projects purely on the basis of its supposedly poor reputation in satisfying the client's needs.

However, for at least the past three decades, although some confusion still exists among inexperienced clients, the term design and build has almost been unanimously interpreted and defined as being:

> An arrangement where one contracting organisation takes sole responsibility, normally on a lump sum fixed price basis, for the bespoke design and construction of a client's project.

This definition contains three elements which are fundamental characteristics of this system, i.e. the responsibility for design and construction lies with one organisation, reimbursement is generally by means of a fixed price lump sum and the project is designed and built specifically to meet the needs of the client.

Genesis

It could be argued that the design and build method is probably the oldest proven procurement system still in use in the UK as, until the middle of the eighteenth century, first the client, then the architect and finally the master builder were, in turn, solely responsible for both the design and construction aspects of most of the building projects implemented during this period of history.

At about this time, the complete separation of design and construction, which had begun during the Renaissance with the emergence of the architectural profession, finally occurred, and the approach that we know of today as the conventional procurement system became the main method used to implement projects.

As this latter method maintained its dominance over all the other systems until the late 1960s, or early 1970s, it is not surprising that design and build only began to emerge from its period of dormancy after the Second World War, and even then only initially to answer the needs of the ambitious targets set by government for the public housing sector.

Without contractor-designed housing systems, the high housing output figures of the post-war years would not have been achieved, and the credibility and viability of a procurement procedure whereby the contractor acts as both designer and constructor would probably not have been established so quickly.

In parallel with the re-emergence of the design and build principle as a tool to meet local authority housing targets in the UK, greater use of this system for industrial and commercial projects was being made in the USA and, gradually, following the lead from across the Atlantic, private-sector clients in this country began to adopt the integrated approach being marketed by contractors.

The overheating of some sections of the national economy in the early 1960s resulting in heavy demands upon the construction industry and a shortage of construction resources, coupled with claims by contractors of greater efficiency and lower cost when using this method, led to the increased use of contractor-designed 'system' building by the public sector in both the housing and non-domestic sections of the market.

Although the use of contractor-designed systems declined in the late 1960s and early 1970s, contractors had by then amply demonstrated their ability, particularly in domestic building, to manage large integrated projects and achieve savings in time; however, little evidence exists that the direct cost savings forecast were ever achieved.

However, the state of the national and international economies again produced an opportunity for contractors to satisfy the growth in interest among clients in finding alternative ways of procuring projects that occurred as a result of the 1973–1974 oil crisis, when a dramatic increase in borrowing and inflation rates emphasised the need to ensure that projects were both speedily commenced and completed in order to minimise the associated financial risk.

At the same time, client dissatisfaction with the performance of conventional methods of building procurement meant that any method where there was single-point responsibility, where the integration of design and construction could lead to savings in time and where fixed price lump sum tenders could be obtained was extremely attractive, and these heavily contractor-marketed characteristics ensured the growth in use of the design and build system and in turn produced one of the most significant trends in construction procurement in recent years.

Share of the market

Efforts to identify accurately how much of the annual non-domestic building workload is carried out using the design and build system are thwarted by the probable use of incorrect terminology in such surveys as are available and the resulting difficulty in determining whether the results refer to the main system or its variants.

Of more importance, however, if one is trying to establish which category of client uses which system, is the fact that different clients use the same procurement system to a greater or lesser extent dependent upon the nature of their business and in some cases the type of project they are implementing at the time.

This phenomenon is best illustrated by the results of the two surveys undertaken by NEDO in 1983 [1] and 1988 [2], when it was established that industrial clients carried out more than one-third of their building projects using the design and build system whereas commercial clients only used this system for less than one-quarter of their schemes.

Most other surveys of the level of use of the various available procurement systems have not differentiated between categories of clients, or their project typology, and thus the results can only be used to identify in general terms the share of the market captured by this procurement system. Two of these surveys are worthy of examination.

A study carried out by Moore [3] found that (1) 24 per cent of the annual workload of the sample of thirty-eight contractors examined was executed using the design and build system and that (2) all of the contractor participants were confident of increasing their involvement in this method in the immediate future.

This confidence was proved to be well founded when between 1984 and 1998 the Royal Institution of Chartered Surveyors [4] carried out biennial surveys to establish the levels of use of building forms of contract and procurement systems by quantity surveyors in private practice, local authorities, housing associations and central government offices. While admitting to the probability that the surveys understate the use of the design and build system, as a result of the nature of the respondents, the results showed that the value of contracts using the system had increased from 5.1 per cent in 1984 to 41.4 per cent in 1998.

The 1996 design and build survey carried out by the University of Reading [5] established that the market share, by value, of the commercial, industrial and housing sectors in 1995 was 32 per cent, although its authors admit that the survey was biased to include, in the eventuality, 85 per cent of respondents who had experience of using the design and build system.

A survey of the use of design and build worldwide described in the Australian journal *Building Economist* [6] published at the end of 1997 found that the frequency of use of the method in Great Britain was 30 per cent in the private sector and only 10 per cent in the public domain and suggested that the use of the system was decreasing.

However, in the same year Boudjabeur [7] established that design and build was the fastest growing procurement system in the UK and suggested that it was likely to account for over 50 per cent of the total construction workload by the year 2000. A number of authorities were quoted to support estimates that the method was, at that time, used in 35 per cent of non-industrial and non-housing projects, with this figure increasing to 45 per cent when industrial and housing work was included.

It can thus be concluded that, although the level of use of the system can vary considerably dependent on the category of client and the type of project commissioned, the share of the market is substantial, has increased dramatically over the past decade and a half and is likely to continue to do so for the foreseeable future although perhaps at a much slower rate than previously.

The process

With the contracting organisation taking sole responsibility for both design and construction, the contractual and functional relationships among the employer, consultants and contractor are simplified when compared with most other procurement methods, with communications being reduced, in theory at least, to a single channel. Figure 5.1 illustrates the relationships between the various members of the project team.

It follows that once the decision has been taken to adopt this method of procurement the process should be a simple one consisting of: the preparation of the employer's requirements; the obtaining of tenders; the evaluation of the submissions on the basis of design, specification and price; the acceptance of the most appropriate tender; and, finally, the implementation and completion of the project.

This simplicity, however, is deceptive, and a detailed examination of the process reveals the fact that the various constituent parts are, in themselves, relatively complex and contain a number of potential pitfalls for the unwary and inexperienced client.

Employer's requirements

The purpose of the employer's requirements is to provide the tenderers with sufficient information, in the form of a brief, to enable their proposals and bids to be formulated without difficulty. The brief therefore needs to be

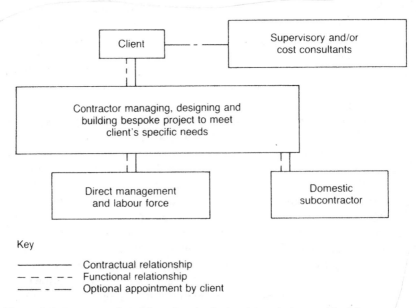

Figure 5.1 Contractual and functional relationships, design and build

clearly presented and sufficiently comprehensive so as to leave the tenderers with no doubt as to the precise wishes of the employer while, at the same time, giving them the freedom to make use of their design, technical and managerial expertise as well as their own particular available resources.

In order to achieve this goal, the employer's requirements should theoretically be drafted only in terms of the project's required performance criteria. However, in practice, the clients needs are, in order to achieve clarity and avoid misunderstandings, more likely to be stated in both performance and prescriptive terms.

This should enable tenderers to be sufficiently aware of the client's requirements in order to ensure the submission of best value bids and easily comparable proposals; however, research has shown that clients are often unable to provide a clear enough brief and that the lack of sufficient information as to their requirements is an inhibiting factor in arriving at a fair and valid assessment of the contractor's bids and the client's ability to make a choice between them. Clients, however, have a wide range of options in formulating the brief, with anything from a short description of the accommodation required to a full scheme design being appropriate, depending upon the nature of the project [8].

It will thus be appreciated that the amount of influence that the client has on the contractor's proposals, and the associated bid, depends upon the employer's requirements being comprehensive and unambiguous as well as the extent to which the client is prepared to elaborate on and clarify his/her requirements before tenders are submitted.

In accepting the responsibility for interpreting the client's requirements correctly, and bidding on a lump sum basis, the contractor is assuming a much higher risk than when he/she is provided with more comprehensive documentation when tendering for projects using other methods of procurement. The burden is increased if the client's needs are not adequately identified and described, and bids may well then include a premium, in the form of an undisclosed contingency sum, resulting in unnecessarily inflated tenders.

As with all briefing documents, achieving the correct composition of the employer's requirements is critical to obtaining the most advantageous design proposals and bids and the overall success of the project.

Obtaining tenders

Once the employer's requirements have been formulated, or even partially completed, the exercise of obtaining tenders can be commenced. This element of the process consists of three basic operations: selecting the contractors who will be asked to submit proposals; deciding upon the tendering procedure; and choosing the form of contract.

Research [9] has suggested that design and build contractors organise their activities in three different ways:

1   Pure design and build – here, the contractor strives for a complete and self-contained unit where all the necessary design and construction expertise resides within one organisation that has sufficient resources to complete any task that arises.

Such firms, it is maintained, rarely undertake anything other than design and build contracts and usually operate within a particular region or, more likely, a number of discrete market sectors.

In such organisations, all aspects of design and construction have the capacity to be highly integrated, and there is a wealth of experience and site feedback which can be utilised for the management of future projects.

2   Integrated design and build – in this form, a core of designers and project managers exists within the organisation, but this type of contractor is prepared to buy in design expertise whenever necessary.

The design and construction teams may well be separate entities within a group, and both design and build and conventional tendered work may be undertaken.

Although more effort is needed to integrate the internal and external members of the design and build team, in-house project managers are employed to co-ordinate these functions.

3   Fragmented design and build – many contractors, both large and small, and including national builders, operate a fragmented approach to design and build projects, whereby external design consultants are appointed and co-ordinated by in-house project managers whose other main task is to take and refine client briefs.

Under this regime, many of the integration and co-ordination problems of the traditional approach are likely to manifest themselves along with some role ambiguity among the professions as they come to terms with the builder as leader of the design and construction team.

It is suggested that the majority of medium and small projects are undertaken by contractors from this category.

Clients and their advisers need to be aware of these categories, and their individual characteristics, when selecting contractors who will be asked to submit proposals. Problems can arise if the chosen organisation is not compatible with the client's operational needs and management structure.

Research [10] maintains that costs, and therefore tender prices, are likely to increase as the choice of contractor moves from fragmented design build through the integrated method to the pure form. It is suggested that the increase in costs results from the reduction in the amount of design work subcontracted to external consultants when using the integrated and pure methods and the fact that the 'no job/no fee' arrangement prevalent in the fragmented and, to a lesser extent, integrated form does not apply to the pure arrangement.

On the other hand, the comparatively small number of pure design and build contractors that operate in the UK have amply demonstrated their ability on numerous occasions successfully to manage large, complex projects which have fully met their client's needs, and presumably cost less in the long term despite incurring short-term additional costs at the design and construction stage.

Having decided upon the type of contractor who would be most suitable for the project, it is necessary to ensure that potential tenderers have the appropriate kind of design and build experience relevant to the current project and the necessary financial capacity and the other required resources. The National Joint Consultative Committee for Building (NJCCB) [11] has produced a questionnaire by means of which the potential tenderers can be tested for suitability.

Other standard selection procedures, such as interviews, obtaining references from previous clients and visits to completed or current projects, can be used as further means of, or aids to, selecting the most appropriate organisations.

The main alternative methods of obtaining tenders are limited to single- and two-stage tendering and negotiation. The choice of procedure will depend on the extent to which the employer wishes to negotiate subsequent to the submission of the tender and the complexity of the project, although if an element of competition is required the option is obviously restricted to the single- and two-stage methods.

The single-stage procedure requires the selected contractors, usually three or four in number, to make only one submission to the client, with the preferred proposal forming the basis for the design and build contract with detailed design work, and the obtaining of the necessary approvals, commencing immediately the contract has been let.

With the two-stage method, up to six contractors can be invited to submit preliminary proposals in the form of outline designs and budget costs. The favoured scheme is selected and the design is developed to an advanced stage and the budget cost converted into a firm bid, which, if acceptable, forms the basis of a design and build contract with work being implemented once the contract has been let.

The increased use of resources at bid preparation stage by design and build contractors when compared with other forms of tendering can be a serious problem, and it has often been recommended that reimbursement of second-stage design costs should be made if the project is abandoned. This recommendation does not apply to the negotiated alternative as this method generally restricts the negotiation to one contractor, with the design and contract sum evolving during discussion between the two parties until the client is satisfied that both elements of the proposal meet his/her requirements and the cost is within the budget.

It is necessary when inviting tenders to advise the bidders which form of contract will be used once the most appropriate submission has been accepted; there are many such forms of contract available, for example:

1   Joint Contracts Tribunal (JCT) Standard Form of Building Contract with Contractor's Design, 1998.
2   Association of Consultant Architects (ACA) Form of Building Agreement, 1982.
3   Government Contract (GC)/Works/1: Edition Three, Single Stage Design and Build Version, 1996.
4   Institution of Civil Engineers (ICE) Design and Construct Form of Contract, 1992.
5   Engineering and Construction Contract (The New Engineering Form of Contract), Second Edition, 1995.
6   Client-drafted forms.
7   Contractor-drafted forms.

It is generally accepted that the use of nationally agreed standard forms of contract provide a stable framework within which the client and the building team can operate with the minimum of difficulty and confrontation.

There are some dissenting voices to this approach on the grounds that contracts should be tailored to suit the individual circumstances of each project and that, in addition, the standard forms lack the strength required to safeguard adequately the client's interests.

However, the latest Royal Institution of Chartered Surveyors (RICS) survey of the use of forms of contract shows that the most commonly used form of contract for design and build is the JCT Contract with Contractor's Design, 1981 (the predecessor to the 1998 edition), with the 1992 ICE Design and Construct contract being the next most popular. Especially drafted employer, consultant or contractor written forms only account for approximately 3.5 per cent by value of the surveyed projects, thus apparently confirming that the majority of the industry's clients, practitioners and contractors believe that standard forms are the most suitable for use with this procurement system.

The evaluation of submissions

The evaluation of design and build proposals and tenders has already been identified as an area where problems are often experienced because of the difficulty of communicating the client's detailed requirements by means of a single written brief, i.e. without the assistance of drawings or a bill of quantities, which can lead to different interpretations by the tenderers.

One suggested method of mitigating this difficulty is for the client, after the initial appraisal has been made, to interview each tenderer's design and management team in order to discuss the individual schemes so as to obtain a better understanding of each proposal. Some clients see this approach as being unethical, and contrary to the guidance given in the NJCCB's Code of Procedure on the grounds that discussions with individual tenderers might result in breaches of confidentiality, and believe that such discussions should be limited to the lowest tenderer.

Advocates of the approach rebut this suggestion by pointing out that any matters of common interest, or changes in the client's requirements, resulting from such discussions can, and should, be communicated to the other tenderers so that they may take the appropriate action. They also point out that, anyway, design and build tenders are not judged solely on the criteria of cost.

Once this exercise has been completed and the client is in possession of all of the information needed to evaluate each proposal, tenders should be ranked in a systematic way in accordance with the attributes of each bid and the client's needs.

The examination and ranking of bids will be made easier by the contract sum analysis: a breakdown that is provided by each tenderer of the lump sum tender into individual sums allocated to predetermined elements or areas of the project. Not only will this analysis enable the client to compare bids and identify any anomalies in tender pricing but will also be of some assistance in determining the validity of any costs submitted by the contractor for additional works or changes that may be requested during the construction phase. The required breakdown of the tender sum will need to be incorporated within the tender documentation.

Jones [12] has produced a simple method (Table 5.1) for evaluating proposals on the basis of ranking and weighting. The use of such a system, which should be designed to suit the client's criteria, is essential in ensuring that the various proposals are correctly appraised and that the chosen proposal provides value for money and fully satisfies the client's requirements.

The inexperienced client may wonder why he/she cannot 'pick and mix' the best elements of each bid in order to formulate a more attractive proposal, and it needs to be understood that not only would this be contrary to the Code of Procedure, and possibly result in a breach of copyright, but also

Table 5.1 Ranking of design and build proposals

Factors	Weighting	Bid 'A'		Bid 'B'		Bid 'C'	
		Rating	Result	Rating	Result	Rating	Result
Price	20	100	20.00	95	19.00	90	18.00
Design	15	90	13.50	100	15.00	85	12.75
Life-cycle costs	25	85	21.25	95	23.75	100	25.00
Technical experience	3	80	2.40	90	2.70	85	2.55
Managerial resources	10	95	9.50	85	8.50	90	9.00
Financial resources	5	95	4.75	90	4.50	95	4.75
Safety record	3	100	3.00	75	2.25	100	3.00
Punctuality	9	85	7.65	75	6.75	95	8.55
Industrial relations	3	80	2.40	80	2.40	90	2.70
Quality record	7	95	6.65	85	5.95	100	7.00
			91.10		90.80		93.05

could end in a joint action by the unsuccessful tenderers to recover their costs as well as undermining the client's credibility with the local construction industry.

Acceptance of tenders

Once the successful proposal has been chosen, it will be necessary for the client and the design and build organisation to enter into one of the forms of contract that have previously been identified. In this context, it is essential to ensure that the contract documentation incorporates not only the original employer's requirements and contractor's proposals but also any amendments or additions resulting from subsequent negotiations between the two parties.

Implementation of the project

Once the contract has been entered into, the contractor will normally require a period of time before commencing work on site during which to obtain the necessary planning, building regulation and any other statutory approvals; to carry out the detailed design, or sufficient of it to commence work on site and maintain continuity and progress; and to allow for any lengthy lead in time for material delivery.

During this period, the client has to ensure that detailed drawings and specifications, proposed suppliers and subcontractors and samples of, or specifications describing, materials and equipment are submitted for approval and officially authorised by his/her representative/agent.

The single-point responsibility for both design and construction should ensure that improved communications and team work will result from the use of this system, although it must be borne in mind that when using an integrated or fragmented type of contractor these characteristics may not be as evident as when a pure design and build organisation has been appointed.

Although during the construction period the contractor is theoretically responsible for monitoring progress and advising the client of any envisaged delays in completion, it would be a naive employer who does not ensure that independent checks are regularly carried out to ensure that the contractor's progress is in accordance with the agreed programme and that the necessary corrective action is implemented should any operation fall behind.

While the contractor is also responsible, under the contract, for exercising control of the quality of materials and workmanship and for the installation of quality-assurance schemes the client can, and should, make provision for the independent monitoring of the quality of the works although care needs to be taken to ensure that this activity is carried out in strict conformity to the terms and conditions of the contract.

The administration of the project is carried out by an agent appointed by the employer, either from within his/her own organisation or from an external construction consultant. The duties of the appointee can vary, but normally include the monitoring of progress and very often quality, with assistance from a clerk of works, as well as the administration of the contractual and financial aspects of the project.

It is in the employer's best interest to ensure that the arrangements made for reimbursement of the contractor during the currency of the project are such as to ensure that difficulties do not arise through disputes as to the level and timing of any interim or other payments that are due.

The most common payment procedure currently in use is that of the lump sum fixed price method, whereby the contractor undertakes to design and construct the project in accordance with the approved drawings and specification for the sum quoted in his/her tender, which is only adjusted when changes in legislation that could not be envisaged at tender stage are implemented and/or the client instructs variations during the design and construction period.

When using this procedure, and the JCT Form of Contract, interim payments will normally be made, either on a monthly instalment basis against a measured amount of work completed or in accordance with a system of agreed stage payments entailing the agreement and reimbursement of fixed sums when specific progress stages have been reached. When this method is being used, the design and build firm applies for payment and the employer's agent checks the application against the amount of work completed, or the actual against planned progress, and issues an appropriate certificate to the employer confirming the amount of payment required.

Alternatively, the client can appoint an adviser, most usually a quantity surveyor, to examine claims for payment and ensure that value for money is being obtained. It has been suggested that this arrangement is less effective than it would appear in that, unless a number of variations have been requested by the client, the examination of interim claims is so simple as not to require such professional expertise.

Contractors, however, appear to support such an appointment, and indeed often go so far as to recommend this course of action to their clients on the grounds that an independent professional must act in an unbiased capacity to ensure that fair and equitable interim and final payments are made. Cynical observers suggest that such an appointment is mere window-dressing by some contractors in order to achieve an aura of respectability for their commercial operations.

The vexed question of variations has been the subject of much discussion in the context of the design and build system of procurement, with one school of thought stating that variations are unusual when using this method because all design decisions are made before construction work commences and maintaining that in addition such changes are discouraged by contractors.

The second, more realistic, school suggests that design and build organisations are aware of the need for flexibility to accommodate a client's particular requirements by means of variations and the desirability of providing a sufficiently detailed breakdown of their tender in the contract sum analysis to facilitate the simple evaluation of such additions or omissions.

It is essential, therefore, that clients ensure that such a breakdown is included in both tender and contract documentation and that the number of variations should be minimised, but where they are required their cost and time consequences are agreed before any instructions are issued.

The product

There is a general agreement that the three main benefits resulting from the use of the system are speed, single-point responsibility and savings in cost.

Speed

On the question of speed and completion on time, a number of authoritative studies have found that, to varying degrees, design and build projects were associated with shorter overall project times than when using conventional systems, although the individual design and construction periods were often longer.

The reduction of the overall project period is attributed to the system's ability to overlap the design and construction phases, improved communications between the various members of the project team, the integration of the two basic functions of design and construction and the improvement in buildability and use of contractor's resources resulting from this last characteristic.

The report *Designing and Building a World Class Industry* [5] found that projects using design and build are 50 per cent more likely to be completed on time than those using conventional methods of procurement; construction is 12 per cent faster than when using conventional systems; and total project times are 30 per cent faster than those experienced on conventional projects.

However, despite design and build projects generally having a better record in terms of overall completion times, it has been found that their ability to satisfy clients is more variable than when using conventional systems. It may well be that this characteristic stems from the existence of the three different types of design and build organisations, i.e. pure, integrated and fragmented, as research appears to have demonstrated that the use of the last type in particular is less successful in reducing the time taken for the overall implementation of the work than the first two.

Single-point responsibility

There appears to be general agreement that one of the main benefits to the client when using this system is the advantage of single-point responsibility, with direct contact between the two parties to the contract enabling misunderstandings to be minimised and procedures to be simplified.

Turner [8] goes even further and suggests that the single point of contact is probably the greatest strength of design and build, outweighing the likely savings in time and cost, and the University of Reading's report [5] found that 30 per cent of the clients surveyed considered this characteristic to be their main reason for using the system and 83 per cent considered it to be an important priority.

It has, however, been pointed out [13] that single-point responsibility can only really be achieved if performance criteria are used in the formulation of the client's requirements and the contractor is allowed sufficient freedom to develop his/her own design and is also prepared to accept the responsibility for any post-contract designs that have been produced by the client's consultants and incorporated into the final product.

Cost

On the question of cost, it has been established that the client is interested, on any project, in the early prediction of the total amount he/she will have to pay and the variance between this figure and the final contract sum as well as endeavouring to ensure that value for money is being achieved.

The design and build approach enables the contractor to be more positive about the final cost to the client at an earlier stage, although cost certainty can only be achieved if the employer's requirements are unambiguous and comprehensive and are not subject to alteration during the construction phase [13].

Evidence also exists to support the widely held belief that when using this system the initial and final costs are lower than when using other methods of procurement basically as a result of diminished design costs, the integration of the design and construction elements and the in-built buildability of the detailed design.

This evidence was given further support by the University of Reading's report [5], when it was established that projects procured using this system were a minimum of 13 per cent cheaper than those using more conventional approaches.

But it has also been found that fragmented design builders performed badly, in terms of client satisfaction and on cost–performance criteria, compared with the other two categories of organisation. It has been pointed out that value for money is difficult to assess because of the different methods, designs and services offered by contractors and the limited amount of information usually available at the tender adjudication stage.

Turner [8] has pointed out that unless the client is aware of the lifetime costs associated with the building he/she will be unable to judge the efficiency of the design using information in tenders based solely upon the construction cost.

Functionality and quality

It is interesting to note that most literature about design and build does not identify high levels of functionality or quality as a benefit when using this method of procurement. This result undoubtedly reflects the still prevalent attitude among some clients, and certainly some architects, that design and build is most suitable for simple uncomplicated projects and the belief that the aesthetics and quality of the finished product when using this method is lower than achieved by other systems of procurement.

Current experience does not support this viewpoint as many large, complex and prestigious projects have been, and are presently being, constructed using this method, seemingly confirming the opinion of many that the risk of the client obtaining a crude design solution, or a substandard project, arises only if the employer's requirements are inadequate and the selection of the bidding contractors is not carried out correctly.

Bennett *et al.* [5] found that the system performs better in terms of quality on complex or innovative buildings than on simpler developments. It was also established that only 50 per cent of projects using this method met the client's quality expectations, compared with 60 per cent of conventionally procured projects.

A growing, and expert, body of opinion contends that the system is suitable for the implementation of most types of building, provided that the employer's requirements are carefully and accurately specified. It may well be that, if the system is used with aptitude and skill by competent clients, designers and contractors, most types of new-build projects can now be successfully implemented using this system.

Having said this, there is a remarkable lack of independent authoritative advice available to clients on how to provide a satisfactory brief and generally commission a design and build project, and this deficiency, added to the possible use of an inexperienced fragmented design builder, can produce a dangerous set of circumstances which could result in the client's functional and quality needs being unsatisfied.

5.3 Variants of design and build

The four systems that are now discussed, i.e. *novated design and build*, *package deals*, the *turnkey method* and *develop and construct*, are considered to be the main variants of design and build, although, as with all procurement systems, many other less common variations of the method may be in use within the industry.

Novated design and build

It has become more common over the past decade for clients to appoint consultants to carry out the conceptual design and preparation of design and build tender documentation for their projects and, once the contractor has been appointed, to novate the design team to the successful bidder to carry out the detailed design as the contractor's consultants.

The tender documentation will contain details of the client's consultants and the proposed novation procedure, together with a requirement that the contractor who is eventually awarded the contract will have to accept responsibility for the total design of the project, including the initial work carried out under the client's aegis. In other words, it is as if the consultants, both legally and practically, had always been the contractors' designers. On occasions, details of the conditions of appointment of the novated consultants by the contractor, including the level of fees, have been predetermined by the client, although normally this is a matter for negotiation between the contractor and the design team.

Turner [8] points out that such an arrangement should enable the design of the project to proceed more smoothly from the precontract to the post-contract stage, although it does mean that any precontract negotiations that take place between the bidders and the client will have to be carried out using the services of the contractors' 'temporary' designers, who were responsible for the outline designs contained within the contractors' proposals.

Once the client's consultants have been novated to the contractor, they will no longer be available to provide advice on the detailed design to the client, and it is therefore likely that he/she will need to employ new consultants to examine the final scheme and to confirm or otherwise its suitability. Some clients may be naive enough to think that the novated consultants may still be representing their best interests, despite being employed by the contractor, however this is extremely unlikely as the designers, for both professional and practical reasons, are not likely to wish to put their relationship with the contractor at risk.

Problems can also arise as a result of the contractor being compelled to employ consultants rather than being able to choose his/her own designers with whom he/she may well have an ongoing and successful relationship on this type of project. This kind of forced arrangement may well produce a less than happy team, in which the necessary high level of integration between design and construction which should be inherent in the design and build system is not achieved.

While many successful projects have been carried out using this method of procurement, there appears to be no general consensus among any of the participants as to its suitability for use on the majority of design and build projects. However, the University of Reading's 1996 report [5] found that this variant led to the worst possible outcome for design and build projects.

Package deals

The package deal system is the precursor and parent of design and build proper. As the name suggests, the intention of the original concept was that clients would be able to purchase a total package, virtually off the shelf, to satisfy speedily their building needs at an economical price.

While the idea of being able to buy a suitable building as if one were purchasing any other large consumer article is attractive in theory, in practice the fact that package dealers provide an adapted standard product means that they are unable to satisfy fully the needs and criteria of the majority of clients.

The fundamental difference therefore between the design and build and package deal systems is that the former method provides a bespoke design solution to suit the client's specific requirements whereas the latter uses a proprietary building system in order to produce a scheme which is unlikely to satisfy all of the client's needs.

Figure 5.2 illustrates the system and the various relationships between members of the project team. Provided that the purchaser's requirements are flexible, this method can be an attractive proposition, particularly as the probable reduction in the design, approval and construction stages of the project can lead to savings in time and cost.

The majority of package deal contractors, by their very nature, employ their own in-house designers and can thus be categorised as pure design

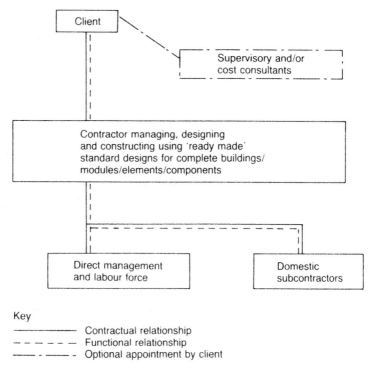

Figure 5.2 Contractual and functional relationships, the package deal

builders; as such, they can be expected to perform well particularly in terms of the speed and time criteria. Some of the products of this method lack aesthetic appeal, but as the potential client is often able to see actual examples of the contractor's product before reaching a decision this potential difficulty can often be avoided.

Another aspect of the use of this particular variant that needs to be carefully examined is that of the proven stability and safety of the design. While many proprietary systems have been tried and tested in use over many years and are often less prone to teething troubles than bespoke designs, some serious failures have occurred in the past and the client would be wise to arrange for this possible problem area to be examined by an independent expert.

In all other respects, the package deal replicates the characteristics of the design and build system, although the forms of contract used with this method are likely to be contractor drafted, rather than any of the nationally recognised standard forms, and great care therefore needs to be taken by clients if this type of document is to be used.

The turnkey method

This system is, as the name implies, a method whereby one organisation, generally a contractor, is responsible for the total project from design through to the point where the key is inserted in the lock, turned and the facility is immediately operational. The responsibility of the contractor is thus, when using this variant, often extended to include the installation and commissioning of the client's process or other equipment and sometimes the identification and purchase of the site, recruitment and training of management and operatives, the arranging of funding for the project and, latterly, under the private finance initiative (PFI) the operation of the project (see Figure 5.3).

The turnkey method was pioneered in the USA in the early 1900s, where it has been extensively used since that time, by the private sector, for the construction of process plants, oil refineries, power stations and other complex production facilities.

Many UK consultants and contractors have been involved in turnkey projects overseas, particularly in developing countries, where the volume of work carried out by this method is higher than in Europe or the USA. The use of the system in this country has been limited, and the amount of work carried out in the industrial and commercial sector appears to be small by comparison to the USA.

However, the introduction of the PFI by the UK government in the early 1980s resurrected a concept that has been established for many years, having been used in the nineteenth century to provide infrastructure such as railways and canals, whereby private finance was used to design, build and operate major public projects such as the Channel Tunnel, the Dartford river crossing and, more recently, various hospitals, prisons, etc.

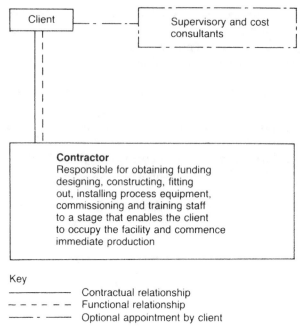

Figure 5.3 Contractual and functional relationships, the turnkey system

Governments throughout the world have now for more than two decades, for political and economic reasons, endeavoured to husband their own resources and cut public capital expenditure by granting the private sector a concession to finance, design, build and operate major public projects which would have in the past been carried out using more traditional methods of procurement.

In order to attract the private sector, various methods of guaranteeing returns on investment, as well as the more mundane, but important, profits on the design, construction and operating elements of the project, have been devised. These arrangements are generally described as concession contracts, but are more commonly known as BOO (build, own and operate), BOT (build, operate and transfer or build, own and transfer), BOOT (build, own, operate and transfer) and DBFO (design, build, finance and operate), the last method being increasingly adopted in the UK [14]. Other less common variants include BRT or BLT (build, rent/lease and transfer) and BTO (build, transfer and operate) [15].

Put simply, these various methods require that the company which has been granted the concession, which is likely to consist of a consortium of contractors and funders each with their own expertise, to design, build, operate and at some specific time transfer the facility to the commissioning client body or even retain it in perpetuity. Figure 5.4 illustrates a typical structure for a BOT project.

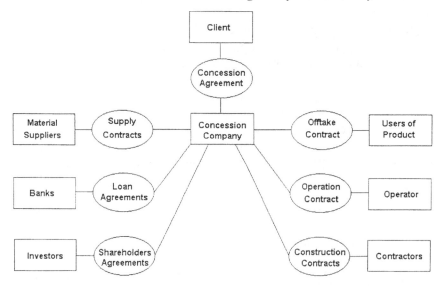

Figure 5.4 Typical structure for a BOT project

During the operating stage, the concessionaire receives income from users of the facility which may be subsidised by the client until the project has reached its forecast user capacity and thus becomes commercially viable.

Although many such arrangements have been implemented throughout the developed world, this method of providing socially orientated projects such as hospitals, prisons has only comparatively recently begun to be used on such developments, but, if forecasts are correct, its use will be increasing in the immediate future despite initial teething problems with the use of the method in the UK.

Apart from the fundamental advantage to the client of being able to take over a fully operational facility, or in the case of PFI schemes reducing public-sector capital expenditure in the short term while establishing a commercially viable development in the long term, the turnkey method from a construction viewpoint echoes all of the characteristics of the design and build system. The forms of contract, however, used with this variant are usually drawn from the process engineering industry rather than from the construction industry.

Develop and construct

When using this system, the client's consultant is provided with a detailed brief, which he/she may also help to formulate, from which he/she prepares conceptual drawings/sketch designs and a site layout, often including the disposition of individual buildings and their plan forms.

The contractor develops the conceptual design, produces detailed drawings, chooses and specifies materials and submits these proposals with his/her bid in the same way as with design and build proper (see Figure 5.5).

The use of this variant is therefore appropriate where the client desires, or needs, to determine the detailed concept of a project before inviting competitive tenders and yet still requires a single organisation eventually to take responsibility for the detailed design and execution of the project.

It is thus apparent that the main difference between design and build and this variant is the extent to which the design of the project has been developed by the client before inviting tenders. In most cases, the design will be developed at least up to outline planning stage and may, in sensitive planning locations, be taken to the point where full planning approval could be obtained.

This method is most frequently used where the client (1) employs his/her own in-house consultants, (2) sees advantages in using a consistently retained consultant with previous experience of similar types of projects, (3) may wish to limit knowledge of his/her intentions to a trusted few and (4) wishes to minimise the differences, so often experienced at tender stage, among normal, individual design and build submissions.

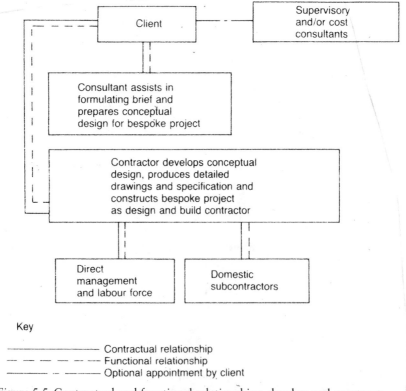

Key

———————————— Contractual relationship
— — — — — —--- Functional relationship
————— — ——— Optional appointment by client

Figure 5.5 Contractual and functional relationships, develop and construct

While it is contended by most authorities that it is the develop and construct contractor's responsibility to ensure the structural sufficiency of the whole design and that the building is fit for its intended purpose, this is a grey area and clients would be well advised to ensure that the question of design responsibility is adequately defined and covered in both the tender and contract documentation for the project.

Once again, all other aspects and characteristics of this variant echo those of the other integrated methods included within this category of systems, with the forms of contract being those previously described for the parent design and build approach.

5.4 Summary

Common characteristics of integrated procurement systems

There are a number of advantages and disadvantages common to all of the methods included within the category of integrated procurement systems.

Advantages

1   The single point of contact between the client and the contractor that is unique to this category of procurement systems means that the client has the advantage of dealing with one single organisation that is responsible for all aspects of the project.
2   Provided that the client's requirements are accurately specified, certainty of final project cost can be achieved, and this cost is usually less than when using other types of procurement systems.
3   The use of integrated procurement systems enables design and construction to be overlapped and should result in improved communications being established between client and contractor. These two characteristics enable shorter, overall project periods to be achieved and project management efficiency to be improved.

Disadvantages

1   If, as often happens, the client's brief is ambiguous and does not communicate his/her precise wishes to the contractor, great difficulty can be experienced in evaluating proposals and tender submissions.
2   The absence of a bill of quantities makes the valuation of variations extremely difficult and restricts the freedom of clients to make changes to the design of the project during the post-contract period.
3   Although well-designed and aesthetically pleasing buildings can be obtained when using this category of procurement system, the client's control over this aspect of the project is less than when using other methods of procurement.

Specific characteristics

Each of the individual systems within the integrated category have specific advantages, disadvantages and characteristics.

Design and build

Disadvantages

The performance of design and build contractors is subject to considerable variation dependent upon whether they are pure, integrated or fragmented organisations.

Levels of technical and managerial competence are likely to be lower as the client's choice moves from the first, through the second to the third type of contractor owing to the difference in capability among organisations specialising in design and build with in-house resources covering all disciplines (pure), a general contractor with partial in-house expertise (integrated) and a medium/small builder in consortium with an out-house design team (fragmented).

Conversely, project costs are likely to increase as the client's choice moves from fragmented through integrated to pure design and build organisations.

Novated design and build

Advantages

The retaining of the same design consultants through all the stages of the process should ensure that design standards are consistently maintained throughout the pre- and post-contract phases of the project.

Disadvantages

There is no guarantee that the novated consultants will be able to establish a good working relationship with the contractor and the forced 'marriage' of the two parties may produce more problems than it solves.

The client may be put to additional expense in appointing new consultants to monitor the design of the project during the post contract stage.

Package deals

Advantages

1   The client is usually able to see actual examples of the package dealers' product in real situations and assess their practical and aesthetic appeal.
2   Many proprietary systems have been tried and tested over a period of years and are thus likely to be free of the initial constructional defects which affect some bespoke projects.

Disadvantages

1   This method uses proprietary building systems to produce schemes which may not satisfy all of the client's needs.
2   Some serious structural failures have occurred among some of these proprietary systems, which have also suffered from other less serious defects as a result of poor design and detailing.

Turnkey method

Advantage

When using this system, the client is able to operate his/her facility and commence operation immediately he/she takes possession of the project.

Disadvantage

The cost to the client of using the turnkey method can be higher than when using other more conventional procurement systems.

Develop and construct

Advantage

This system is useful when the client has his/her own in-house design expertise, regularly uses external designers and sees advantages in retaining them, wishes to restrict the knowledge of his/her intention to build or wants to minimise the difficulties of comparing disparate design and build submissions while at the same time requiring a single organisation to take responsibility for the detailed design and construction of the project.

Disadvantage

Responsibility for the design of the project can be a possible area of dispute when using this system owing to the involvement of both the design consultants and the contractor in this aspect of the project.

   The effect that these characteristics have on the selection of the most appropriate procurement system for a specific project is discussed in detail in a later chapter, but as a general guide it can be safely stated that most new-build industrial and commercial projects – where the client's requirements can be unambiguously and comprehensively delineated, where they remain constant during the currency of the project and where they are not required to be of high architectural merit – could be procured by one or more of the integrated procurement systems that have been described.

This is not to say that aesthetically challenging and high-quality projects cannot be successfully achieved using these systems but rather that clients should be aware that in order to implement such projects they will need to be extremely disciplined in the way the project is managed. The establishment in 1999 of the Design and Build Foundation – a body intended to regulate design and build operators by registration, education and other means – may well help to widen the successful use of the system as well as raise standards.

References

1   Building Economic Development Committee (1983) *Faster Building for Industry,* London: National Economic Development Office.
2   Building Economic Development Committee (1988) *Faster Building for Commerce,* London: National Economic Development Office.
3   Moore, R.F. (1984) *Response to Change – The Development of Non-traditional Forms of Contracting,* occasional paper no. 31, Ascot: Chartered Institute of Building.
4   Davis, Langdon and Everest (2000) *Contracts in Use: A Survey of Building Contracts in Use During 1998,* London: Royal Institution of Chartered Surveyors.
5   Bennett, J., Pothecary, E. and Robinson, G. (1996) *Designing and Building a World Class Industry,* Reading: Centre for Strategic Studies in Construction.
6   Anonymous (1997) 'Design–build around the world', Australian Institute of Quantity Surveyors, *The Building Economist* December, 32–33.
7   Boudjabeur, S. (1997) 'Design and build defined', in *Proceedings of the 13th Annual Conference and Annual General Meeting,* Cambridge: Association of Researchers in Construction Management.
8   Turner, D. (1995) *Design and Build Contract Practice,* 2nd edn, Harlow: Longman.
9   Rowlinson, S. (1987) *Design Build – Its Development and Present Status,* occasional paper no. 36, Ascot: Chartered Institute of Building.
10   Rowlinson, S. (1986) 'An analysis of the performance of design build contracting in comparison with the traditional approach', unpublished PhD thesis, Brunel University.
11   National Joint Consultative Committee for Building (1995) *Code of Procedure for Selective Tendering for Design and Build,* London: Royal Institute of British Architects Publications.
12   Jones, G.P. (1984) *A New Approach to the JCT Design and Build Contract,* London: Construction Press.
13   Mosey, D. (1998) *Design and Build in Action,* Oxford: Chandos Publishing.
14   Walker, C. and Smith, A.J. (1995) *Privatised Infrastructure: The Build Operate Transfer Approach,* London: Thomas Telford.
15   Haley, G. (1996) *A–Z of BOOT, How to Create Successful Structures for BOOT Projects,* London: IFR Publishing.

6 Management-orientated procurement systems

6.1 Introduction

The last three decades have seen a substantial increase in the use of management-orientated procurement systems in the UK, mainly as a result of clients, particularly those in the commercial sector, demanding earlier commencement and completion times than could be achieved when using conventional methods, more control over project costs and higher standards of functionality and quality than were being obtained by other means of building procurement.

It is suggested that the introduction of management contracting was the result of changes which had occurred since the early 1960s in respect of:

1   the diversity, complexity and standardisation of building techniques;
2   the growing prominence of the subcontractor;
3   the growth in size of projects, demand for tighter time and cost targets and for a more unified and purposeful management of the total process.

The systems that have been included within this category exhibit the characteristics common to all management-orientated systems whereby the contractor is elevated to the status of a consultant and special emphasis is placed on the integration of the management of both design and construction. This trio of methods consists of *management contracting, construction management* and *design and manage.*

All of the systems that are now described should, in theory, enable the project to be more readily effected in accordance with the client's needs, especially accelerated commencement and completion, than would be the case if the work was carried out using more conventional procurement systems.

6.2 Management contracting

Definition

It should be said from the outset that this procurement system is examined from the viewpoint of the procedures laid down in the documentation associated with the 1987 and recently revised 1998 editions of the Joint Contracts Tribunal (JCT) Standard Form of Management Contract and does not therefore reflect any of the very varied non-standard methods of operation used before that date, although reference may be made to these in the text.

Reference must also be made to fee contracting at this stage. This method is where, in exchange for a fee, the contractor agrees to carry out building works at cost while at the same time providing some limited management expertise to the client and his/her design team.

This system of procurement has a long history of specialist use in the UK dating from the 1920s, was the forerunner of management contracting proper and is in reality an integral part of the evolution of the method in current use. This and the fact that the characteristics of the original fee contracting approach are basically those of the cost-plus method that is described elsewhere has led to the decision not to treat this arrangement as a separate entity but as part of management contracting itself.

The main characteristics of the management contracting system are:

1   The contractor is appointed on a professional basis as an equal member of the design team providing construction expertise.
2   Reimbursement is on the basis of a lump sum or percentage fee for management services plus the prime cost of construction.
3   The actual construction is carried out by works or package contractors who are employed, co-ordinated and administered by the management contractor.

Many definitions of the method exist, but all contain, in part at least, these fundamental features. For the purposes of this guide, the following definition has been adopted:

> Management contracting is a process whereby an organisation, normally construction based, is appointed to the professional team during the initial stages of a project to provide construction-management expertise under the direction of the contract administrator.
>
> The management contractor employs and manages works contractors who carry out the actual construction of the project and he/she is reimbursed by means of a fee for his/her management services and payment of the actual prime cost of the construction.

Genesis

While the concept of management contracting appears to have originated in Sweden [1], the system was also developed in the USA as a result of the General Services Administration (GSA) and the Public Building Service (PBS) requiring phased construction and greater consultant's and contractor's co-operation on the civilian federal building projects for which these two organisations were responsible.

According to Heery and Davis [2], until the late 1960s, management contracting – known incidentally in the USA at that time as professional construction management (PCM) – was rather informally used but, as the cost of construction increased during the early 1970s and delayed projects became more frequent, PCM began to be used more frequently.

In the mid-1970s, the GSA restricted construction management firms to acting solely in a consultant capacity in order to preclude them from undertaking any of the direct works. At the end of the decade, the same powerful organisation further limited the use of the method as a result of difficulties experienced in ensuring that there was enough incentive to perform, problems regarding liability and the need for a firm price tender before commencement on site – the wheel had turned full circle!

Despite this, there is no suggestion that the management contracting system has not continued to be used, where appropriate, in the USA, although Flanagan [3] and others have suggested a reduction in its use had occurred in the late 1980s and that this trend was likely to continue into the 1990s.

Although fee contracting was used as early as 1928 in the UK by Bovis, it was not until 1969 that pure management contracting gained recognition when the Horizon project, a large complex cigarette manufacturing factory for John Player in Nottingham, was built using this method [4].

Whereas in the USA the growth in PCM was led by architectural and engineering consultancy practices, in the UK the major contracting companies were the first to offer management contracting services to their clients, with the number of contractors increasing dramatically in the late 1970s and 1980s.

Before, and for most of, the 1970s the large majority of, if not all, management contracting projects emanated from the private sector, and it was not until the end of the decade that the method began slowly to be accepted and used by public clients who had succeeded in overcoming the difficulties associated with the need to ensure public accountability. After initial reluctance, the use of the method increased, mainly as a result of a perceived success rate and good track record, with both local and central government carrying out major projects using the system.

In the late 1980s and up to the mid-1990s, growth in the use of the management contracting system declined, when measured by both value and number of contracts. Indeed, it has been suggested that it was losing out in favour of construction management, although the most reliable statistical data do not support this suggestion. Notwithstanding this possibility, the launch of the long-awaited JCT management contract in 1987 ensured that the method joined the other established and officially approved and accepted procurement systems on offer to the industry's clients.

However, it has been noticeable that, since the middle of 1989, a number of influential clients have publicly declared their intention not to use the management contracting system on any future projects owing to the difficulties they have experienced when using this method for past and current schemes. This announcement was supported by the 'Report and guidance' issued by the Construction Management Forum [5] in 1991, which compared the experienced client's perception of the practical use of management contracting and construction management and concluded that the latter method of procurement was preferred.

The problems that have led to clients' apparent disenchantment with this method are those of delays caused by trade/package contractors over which the client has no control along with the additional costs incurred when using this system, the lack of a contract sum and the difficulty experienced in achieving reasonable quality standards.

Despite these concerns and opinions, management contracting continues to be used more, when measured by value or number of contracts, than construction management and is considered by many authorities [6] to be as suitable for use on appropriate projects as the latter method.

Share of the market

Over the last three decades, management contracting has become well established in the UK as an acceptable alternative to other, more conventional methods of procuring buildings of all types. Naoum's 1988 survey [7] of 170 projects that had been implemented using the management contracting system confirmed that, whereas the majority of the projects identified were in the commercial and industrial sectors, some residential, health and other types of buildings had also been carried out using this method.

Figures 6.1 and 6.2 from the same survey show that both new-build and refurbishment projects and privately and publicly commissioned schemes have been successfully carried out using this method of procurement.

The latter part of the 1970s and most of the 1980s saw a substantial increase in both the numbers of management contractors and the annual value of work carried out using management contracting. Since that period, the survey carried out by the Royal Institution of Chartered Surveyors (RICS) of 'contracts in use' [8] shows that both the percentage by value and the number of contracts using management contracting decreased during the early and mid-1990s but showed an increase in use in the 1998 survey.

A number of authorities in the recent past have hypothesised that the trend towards the increased use of management contracting in the 1980s might not be so inevitable in the future and it would appear that, as a result of the general disillusionment with the system expressed by some major clients of the industry and subcontracting organisations, during the 1990s, at least, this forecast has proved to be correct.

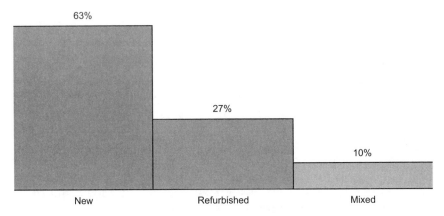

Figure 6.1 Percentage of management contracting by type of construction

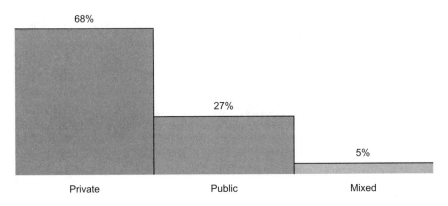

Figure 6.2 Percentage of management contracting by sector

The process

Figure 6.3 illustrates how the system operates functionally and contractually.

The management contract arrangement does not fit neatly into the conventional pattern of precontract and post-contract stages, and Dearle and Henderson [9] have identified three distinct periods:

1   the period before the appointment of the management contractor;
2   the preconstruction period;
3   the construction period.

The question of timing of the appointment of the management contractor should be aired at this stage as there are two schools of thought regarding this matter. A number of respected sources suggest that the management contractor should be engaged in advance of the architect and other consultants and made responsible for their appointment and the reimbursement of the design team. The more usual arrangement, however, and that laid down in the JCT Practice Note for the 1987 Management Contract [10], is for the client directly to appoint his/her consultants in advance of the engagement of the management contractor and to be responsible for their reimbursement and management.

The three stages of the process are now described and discussed based upon the procedures incorporated within the JCT standard form of contract for management contracting.

Preappointment of the management contractor

This period will encompass many of the activities usually associated with the conventional precontract stage, i.e. carrying out of a feasibility study, formulation of the brief, etc.; however, they will be carried out in this case on an accelerated time-scale.

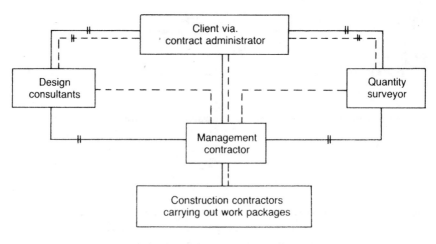

Key

———————   Contractual relationship
– – – – –   Functional relationship
——#——   Alternative relationship

Figure 6.3 Functional and contractual relationships, management contracting

During the early part of this stage, the employer appoints his/her contract administrator, an architect, a quantity surveyor and any other professional advisers that are considered necessary – these consultants form the design team, which initially assists in the preparation of the brief and prepares drawings and a specification that describe, in general terms, the scope of the project.

Once this documentation has been prepared, the employer, with advice from appropriate members of the professional team, invites tenders from management contractors using one of the methods of selection and appointment that have been devised for this purpose. All of the various methods are likely to contain the following fundamental elements:

1 A list is formulated of management contractors, probably of some ten to fifteen in number.
2 These contractors are invited to submit a written reaction to a brief description of the project including the time-scale and estimated cost of the works.
3 On the basis of their initial response, three to five management contractors are selected to proceed to the next stage and are issued with a questionnaire, the answers to which will provide an accurate profile of their organisation and experience of management contracting. They are also provided with detailed information about the project, its design, programme and cost aims, and invited to submit their proposals for managing the project and, separately, their fee bid – expressed as a percentage of the total project cost – for implementing these proposals.
4 It is also usual for the contractor to be asked to state the reimbursement that he/she will require for providing the necessary management services during the preconstruction period.

The evaluation of the contractors should be made on the basis of their written proposals and verbal presentation and should not be determined solely by the level of the individual fee proposals. Great emphasis needs to be placed upon the importance of the way in which the management contractors propose to manage the project and the content and manner of their project team's live presentation of this aspect of the selection process.

The criteria for the selection of the successful management contractor should include:

1 Experience of management contracting and a previous record of success in performing this and the conventional general contractor's role on projects of a similar size, complexity and type and completing work on time and within budget.
2 A detailed understanding of all aspects of the scheme and the evident ability to match his/her management team to the needs of the project –

this is particularly so in the case of the project manager, whose competence and experience will be crucial to the success of the project.

3   Adequate financial resources, with sufficient capacity for the bonding and insurance requirements of the project.
4   Sufficient standing within the industry to ensure that any difficulties with suppliers or trade/package contractors are dealt with expeditiously.
5   The ability of both the head office organisation and the site project team to co-operate with the client and his/her consultants throughout the life of the project.
6   The level of the management fee.

An examination of management contracting [6] revealed that the client's choice of management contractor not only will depend on the quality of individual bids but also will reflect the proposed fee, the contractor's perceived understanding of the project and his/her experience and capability. As the client's perception of the latter qualities may well be based upon the presenters of the individual bids, he/she may well specify that these individuals from the winning bidder must be part of the management contractor's project team.

The aim of the selection procedure is not to achieve the lowest fee and other charges but to select the management contractor with the right experience, resources and managerial skill, and the client would be well advised to use one of the various assessment techniques that are now available in order accurately to evaluate the subjective variables that will be present in each submission.

Once the management contractor has been selected, he/she and the employer enter into a form of contract – the JCT 1998 Management Contract now being accepted as the industry standard – and the second phase of the project commences.

Preconstruction period

During this period, the management contractor provides certain services to the client's design team. These will normally include [10]:

1   the preparation of the overall project programme;
2   the preparation of material and component delivery schedules and the identification of those items which require advance ordering;
3   advising on the 'buildability' and practical implications of the proposed design and specifications;
4   the establishment of agreed construction methods, particularly those affecting the design;
5   the preparation of a detailed construction programme;
6   advising on the provision and planning of common services and site facilities;

7 advising on the breakdown of the project into suitable packages for trade and/or works contractors;

8 preparing a list of potential tenderers for the works or trade contracts and investigating their capabilities and financial standing;

9 assisting in the preparation of tender documents and obtaining tenders from approved trade/works contractors and suppliers;

10 the preparation, in consultation with the client's consultants, of the documentation necessary to ensure the efficient placing of the proposed trade/works/suppliers contracts or orders.

A number of the listed responsibilities impinge on those that are normally included in the duties of the design team, which, while remaining responsible for the design of the building, must be prepared for a dialogue with the management contractor, for example, on methods of construction and buildability in order to reach an agreed design which best meets the client's needs.

During this period, the management contractor has other functions, which must be carried out before the end of the preconstruction phase, including the agreement of the contract cost plan/estimate of prime cost with the cost consultant.

Other matters relate to the agreement of the services to be provided by the management contractor: the terms of the joint policy for insuring the project; the identification of the management contractor's resident site manager; and the agreement of the common site facilities and services that will be provided or secured by the management contractor.

When the contract administrator is satisfied that all the preparatory work has reached a stage when it will be practicable to commence the construction of the project, he/she notifies the client, who then has 14 days in which to decide whether the project will proceed or not.

If the project is aborted at this stage, the management contractor and design team are reimbursed their costs in accordance with their respective contracts. In the case of the contractor, he/she is paid the preconstruction period fee, which will have been agreed and stated in the contract.

If, however, the management contractor is notified that he/she is to proceed, he/she is given possession of the project site in accordance with the previously agreed date for possession and thus becomes responsible for securing completion of the project on, or before, the date for completion or any extended date fixed under the conditions.

Construction period

The management contractor does not carry out any of the construction work himself/herself. The actual work is divided into a series of separate packages, generally for different trade or functional elements of the building, and the management contractor usually enters into standard JCT contracts with works contractors for the implementation of these packages.

The duties of the management contractor during this period can be summarised as setting out, managing, organising and supervising the implementation and completion of the project using the services of his/her trade/works contractors.

The selection of tenderers, the establishment of the conditions of the works contracts, the obtaining and scrutiny of works contract tenders and recommendations for the acceptance of the most favourable offer are all matters which are dealt with jointly by the client, the appropriate consultants and the management contractor.

The management contractor is precluded from carrying out any of the trade/works packages, either himself/herself or through any associated or subsidiary company, and from profiting from the construction activities so that he/she can avoid any conflict of commercial interest between the client and himself/herself and retain his/her consultant status.

Payment is therefore made to the management contractor during the construction period by means of interim certificates based upon the prime cost of the work so far completed. At the same time, directions are given for the amounts that are to be paid to the individual works contractors. The certificates will also include the reimbursement of an instalment of the construction period management fee and the cost of the management contractor's directly employed resources. Retention is deductible on all elements of the prime cost, including the management contractor's fee.

The monitoring of the cost of the project against the contract cost plan is carried out on a continuous and regular basis by the architect and quantity surveyor, with the assistance of the management contractor, in order that accurate and up-to-date information on committed and future costs is always readily available.

The consecutive receipt of tenders for the works contracts enables the consultants to maintain the contract cost plan by the redesign of later packages if this is judged to be necessary in the light of the result of earlier tenders. The management contractor's own costs should be monitored against an agreed figure contained within the contract cost plan.

The architect has the power to issue variations on the project, generally described as 'project changes', and also for the works contracts, described as 'works contracts variations'. He/she can also order a postponement and issue instructions to accelerate the progress of the project, or alter its sequence, or timing.

As with most other procurement systems, variations to the original brief are to be avoided if at all possible, but, should such variations be necessary, it is essential that the management contractor and other appropriate consultants assess the effect of the proposed changes on programme and cost before a decision to proceed is made.

The formulation of the construction programme necessary to achieve the completion date is the responsibility of the management contractor, and although extensions of time can be granted under the JCT form of management contract these do not necessarily affect the project completion date. This approach is intended to ensure that the consultants and the management contractor concentrate on completing the project on time rather than on formulating reasons for obtaining an extension of time.

However, the JCT management contract has been designed deliberately to provide a low-risk environment for the management contractor, which results in his/her liabilities being limited – in the same way as the other members of the professional team – to responsibility for any negligence in the performance of his/her management function. As a result of this approach, the management contractor is given relief from the imposition of ascertained and liquidated damages for any overrun on the project which is not covered by an extension of time and which has been caused by reasons outside his/her reasonable control, or indeed for delays caused by any of the works contractors. However, recent case law has undermined this principle, and the current legal position appears to be somewhat unclear as to the extent of the management contractor's liability for non-performance by the works contractors or others.

During the construction period, the management contractor is responsible for supervising the construction operations and ensuring that all of the work is built to the standards originally defined by the design team. This responsibility can be supplemented by the appointment of clerks of work, with the cost being borne by the employer.

Should any of the works contractors be unable or unwilling to remedy any defects identified in the work they have carried out, it may not be possible to recover the cost of rectifying the faults. In line with the philosophy of the management contractor's status as a consultant, the cost of the necessary rectification devolves upon the client, and employers are therefore advised to establish a contingency fund to cover the risk.

Once substantial completion of the works has been effected, handover of the project is carried out in the normal traditional manner, with the management contractor providing the following:

1   A written undertaking that the project conforms to the original or amended brief.
2   A written undertaking that all costs have been paid, that there are no claims, litigation or the like pending and that all works contractors' accounts have been finalised or will be finalised within a maximum period of 12 months.
3   'As constructed' drawings and a comprehensive specification for all aspects of the project.
4   Commissioning and test certificates, servicing and maintenance manuals, and operating instructions for all elements of the building and services including equipment and plant.
5   A written programme of recommended maintenance for the building, services and plant/equipment.

6   A written undertaking confirming that final testing and balancing of the building services will be carried out once the building is fully occupied and operational.

7   Written confirmation of the date of commencement and completion of the defects liability period.

8   A certificate, issued by the architect, confirming that the project is to his/her total satisfaction.

It cannot be sufficiently stressed that the management contract is a high-risk form of procurement for the client, and it is therefore essential that when using this system he/she is completely and continuously involved in the project and that his/her contract administrator/project manager is highly competent and experienced.

The product

The performance of management contracting in relation to other conventional methods of procurement has been well documented in literature and research, so the method is now examined in the context of the three classic clients' criteria of functionality/quality, cost and time.

Function and quality

Provided that the management contractor is appointed during the early stages of the project, and in advance of any decisions being taken regarding the detailed design of the building, he/she should in theory – by advising on buildability, construction methods and techniques, and the economics of the proposed design – be able to make a major contribution to achieving an agreed design which fully meets the client's needs and is capable of being economically and speedily implemented.

However, for this theoretical advantage to become a reality, it is essential that the design team is prepared to enter into an open dialogue with the management contractor on all aspects of the project and to ensure that he/she is well briefed as to the design constraints and the client's requirements so as to avoid unhelpful and impractical suggestions being made.

The client must therefore ensure that his/her consultants are fully aware of, and committed to, the philosophy of the system, are receptive to suggestions from the management contractor on any aspect of the design and are desirous of working as a team in order to find solutions to problems that arise during both the design and construction stages of the project; notwithstanding such co-operation, the design team remains responsible for the design of the project.

While the majority of authorities believe that the management contractor can make a worthwhile contribution during the design phase of the project, some are not convinced of the validity of this claim and maintain that the management contractor's personnel are generally constrained by a lack of experience and understanding of the design function and the need to change from the narrow profit-orientated approach of the general contractor to the wider view that needs to be taken by the client-orientated consultant.

In practice, the amount of assistance that the management contractor can, or is allowed to, give during the design stage is probably limited as a result of him being employed at a late stage in the design process and/or his/her comments and advice not being accepted, or ignored, by the design team [6].

On the question of the quality of the completed project, the general perception appears to be that the quality standards achieved are at least as good as those obtained on conventionally procured projects, and in some cases have been better.

It should, however, be noted that, whereas quality control is the primary responsibility of the management contractor, the architect may also wish to be represented on site in order to monitor this aspect of the project. In these circumstances, there may be some duplication of site supervision, and the costs associated with this activity, as well as the possibility of friction between the management contractor and the designer, which could be to the detriment of the client. This potential problem should therefore be resolved before construction commences on site.

Time

The use of this method is seen by most authorities as being appropriate where fast design and construction periods are required or where an early start, and/or early completion, is needed. On major projects where these criteria must be met and the use of conventional methods of procurement is not appropriate, management-orientated systems have flourished as a result of their perceived ability to fragment the building operations and enable design and construction to overlap.

Sidwell's research [11] into the attitudes of client organisations towards management contracting supports this perception in that 100 per cent of the respondents to a questionnaire comparing this method with the conventional approach were of the opinion that management contracting allows an earlier start to be made on site and resulted in speedier completion of the project; 90 per cent of the respondents also believed that the use of this system resulted in more reliable predictions being made as to the eventual length of the construction period.

Barnes's comparative study [12] of three hospital projects, one using a conventional procurement method and the other two using management contracting, found that 'the pace of construction work on all of the projects did not differ sufficiently to conclude that management contracting has an overwhelming advantage in speeding up construction'. However, the study did establish that the risk of delay was reduced as a result of the system's flexibility and its ability to be able to accommodate and overcome difficulties and changes.

The authors of the Construction Industry Research and Information Association's (CIRIA) special publication no. 81 [6] concluded that savings in time can be achieved from the overlapping of design and construction and from the expedition by the management contractor of the production of the project design. The construction process itself, however, often may not be accelerated as a result of using this method.

Cost

The client's perception of the financial consequences of using this procurement system, as identified in the CCMI's 1985 survey [13] of management contracting, was that it tended to cost more than conventional methods; in addition, the uncertainty regarding the end cost was seen as a distinct disadvantage.

A survey of ten experienced clients carried out by Naoum and Langford [14] found that:

1   All of the respondents believed that management contracting was more profitable to the contractor than to the client.
2   Thirty per cent of the clients surveyed were of the opinion that management contracting involved fewer claims, 40 per cent believed that the method involved the same number of claims as other methods and the remainder did not believe that the method involved fewer claims.
3   Only 20 per cent believed management contracting was cheaper than other methods, 40 per cent thought the method produced the same level of cost and the remainder were of the opinion that it was not cheaper.

In addition, although the authors conceded that there were conflicting opinions about the cost factor, the views expressed appear to confirm the results of the CCMI survey.

CIRIA's 1995 *Planning to Build?* [15] suggests that use of the system may help to minimise costs by improving buildability (and therefore design), by better packaging of activities to enable competitive tenders for all of the work to be obtained and by the phasing of completion to allow phased letting and occupation.

Others have questioned whether management contracting is the most cost-effective way of building, although all have acknowledged that the savings of time that can be achieved may well result in cost savings, or additional income, offsetting the extra construction costs. There is also a suspicion that 'relieved from the constraints of competition, the management contractor's expenditure on site establishment, plant and other preliminaries is lavish – at the client's expense' [11].

Problems that emerged during the late 1980s have resulted in management contracting becoming even less economic in straight financial terms than other procurement approaches. Trade contractors became reluctant to undertake projects at competitive prices because of the onerous contract conditions imposed upon them by management contractors and because of the high level of risk they were being asked to accept. Extra costs had been incurred as the result of the duplication and overlapping of site management and other services by the principal and subsidiary contractors owing to the lack of the economic definition of common services and supervisory responsibilities by the management contractor.

It has, however, been suggested that these were temporary phenomena which have now been dealt with and that the use of the JCT management contract has had a stabilising effect, although this document limits the management contractor's liability to that of professional negligence for his/her advice and thus imposes further financial risk upon the client. Again, this situation may be ameliorated by the consequences of recent decisions of the courts.

The absence of a tender sum when construction commences is seen by many clients as a distinct disadvantage, and to some as reason for not using the system at all. Exponents of the method maintain that the lack of such a safeguard does not necessarily mean that there is less control over the eventual construction cost as strict supervision can be exerted over this aspect of the project. This contention is based upon the fact that each package of work is normally the subject of keen competition between construction contractors, and it is therefore possible to monitor costs closely and, if necessary, adjust later works packages in order to absorb any cost overruns incurred so far, subject, of course, to the client's acceptance of the reduction in scope or specification of the package in question.

The fragmented nature of the construction element of this procurement system allows the final account for the project to be settled progressively during the duration of the works and ensures rapid settlement once the project has been successfully completed.

Other characteristics

In theory, this system enables the client's consultants and the management contractor to become a professional team of equals committed solely to meeting the client's needs. In addition, the emphasis on the management aspect of construction activities should be beneficial and should ensure the effective control of the project.

Specifically, the following characteristics are associated with this method:

1   A high degree of flexibility is built into this method, which enables delays to be overcome or reduced, changes to be absorbed and rescheduling of works packages to be implemented.

2 The fragmented structure of the construction process, i.e. the use of works packages, means that the financial failure of any of the works contractors only has a limited effect on the total project.

3 The management contractor's knowledge and experience should ensure that industrial relations on the project are better than when using more conventional procurement methods.

4 Research and development of new techniques and processes may result from the greater involvement of the contractor during the design stage.

5 The initial philosophy, which has been fundamental to this system since its inception, of the management contractor's liabilities being limited, like all other members of the professional team, to any negligence in the performance of his/her management function means that the client must accept far more risk and responsibility than when using many other procurement systems.

The JCT form of management contract reflects this philosophy, but many of the forms of contract that were used before its publication, and may still be used in the future, do not adopt this approach and force contractors to accept risks more appropriate to conventional contracting.

6 Contrary to the original principle, management contractors have been made by some clients to accept the risks associated with construction rather than just the management of the project. Specifically, clients have made management contractors responsible for construction contractors, time overruns, remedying of works package defects and even design.

These responsibilities bring risks which are similar to those associated with conventional procurement arrangements and place the management contractor in a position where he/she needs to decide whether to accept the unreasonable risks himself/herself, and pass on only the reasonable hazards to his/her works contractors, or whether to ensure that the latter takes all of the risk with the inevitable consequences. The wide use of the JCT form of management contract may have ensured that this deviation from the original philosophy has not irreparably damaged the method.

7 There is little doubt that this procurement system increases the amount of administrative effort required of all of the participants, but particularly the client, and produces a much greater amount of paperwork than many of the other methods.

All of the individual methods of procurement contained within the management-orientated category have a number of characteristics in common and, thus, it is not proposed to reach any conclusions as to the advantages and disadvantages of the management contracting system until all of the members of this group have been considered.

It is, however, apposite to say at this stage, particularly in light of the tribulations that this particular system has experienced in the past, that management contracting is not, as it has often been seen, a panacea for all of the construction industry's ills. But rather, if properly applied, it is a viable alternative to conventional methods, especially where the project is large and/or complex, where there is a need for flexibility and where there is a need for an early start or rapid completion [6].

5.3 Construction management

Definition

The Construction Round Table's business client's guide to using the construction industry, *Thinking About Building* [16], defines this system as the method where the management service is provided by a fee-based consultant, a specialist construction manager or a contractor and where all construction contracts are directly agreed between the client and the trade (package) contractors. It will thus be seen that the fundamental difference between this procurement system and management contracting is that, with this approach, the client enters into a direct contract with the individual works contractors. The construction manager then acts as the employer's agent when dealing with each of the separate contractors. The main characteristics of the system are:

1 The construction manager is appointed as a consultant during the initial stages of the project and has equal status to the members of the design team.
2 Reimbursement is made by means of a lump sum or percentage fee for management services.
3 The physical construction of the project is carried out by works, or package, contractors who are employed by the client and co-ordinated, supervised and administered by the construction manager.

For the purposes of this guide, the construction-management system has therefore been defined as:

The construction manager adopts a consultant role with direct responsi-bility to the client for the overall management of the construction of the project, including liasing with design consultants, to meet agreed objec-tives.

The construction manager is reimbursed by means of a professional fee and all construction is carried out by means of works packages which are the subject of direct contracts between the client and the package contractors.

Genesis

In the USA, architects and engineers began to offer construction-management services to their clients during the l960s. This method has usually been referred to in North America as construction project management (CPM).

It was not until a decade later, during the 1970s, that this method began to be used in the UK in response to demands from clients of the industry for greater certainty about the overall performance of their project, and in particular the need to control the risks of cost and time overruns and detrimental external influences on very large and complex projects.

These problems were further exacerbated when, in the mid-1980s, a depressed and uncertain world economy, coupled with high interest rates, exposed commercial developers to large financial risks and construction management appeared to provide extra management expertise and a control structure which was seen as being necessary to deal with the increased levels of uncertainty arising from these effects.

Since that time, the system has been increasingly used by major developers, particularly on projects in London and the south-east, with apparent success, although often to the detriment of UK consultants and contractors who seem unable to meet the standard of service and cost parameters demanded of, and achieved by, American and other European design and construction organisations.

The use of construction management in the UK by the public sector during its early years was limited, but after a period of experimentation government agencies are now using this procurement system where appropriate and, for the reasons given in the preceding section, are tending to forsake management contracting in favour of the construction-management approach.

Share of the market

The biennial surveys of the level of use of building contracts carried out by the RICS have collected data on the use of construction-management agreements since 1989. The latest survey [8] shows an average level of use of just under 8.5 per cent of all contracts measured by value over the 10-year period up to 1998 and slightly more than 0.5 per cent by numbers of contracts for the same period.

Such other information as is available indicates that the amount of building work carried out using this procurement method has substantially increased, from what must have been a very low level, since the 1970s and confirms the figures published by the RICS that the use of the construction-management system is less common than the utilisation of management contracting, particularly in terms of the number of contracts.

Notwithstanding the lack of further definitive information, sufficient evidence exists from the technical press and anecdotal evidence to confirm the current popularity of construction management, for use on their major projects, among the larger more sophisticated developers and other expert and discerning clients.

The process

There is considerable variation in construction-management procedures and practice and therefore the following discussion concentrates on that process which appears to have become accepted within the industry as good practice.

The functional and contractual operation of the system is shown in Figure 6.4. There is general agreement that the construction manager should be appointed as early as possible in the life of the project, alongside and of equal status to the designer, with both parties dealing directly with the client.

This equal-status situation can only be effective if there is an experienced third party who is directly responsible to the client for co-ordinating and managing the activities of the two organisations. An examination of recent literature describing current construction-management projects reveals that, generally, this co-ordinating role is carried out by the client's own in-house project management personnel and the client is thus more intimately involved with the management of the project than if a more conventional procurement method had been used.

As a consequence of this characteristic, it is unlikely that any but those entrepreneurial clients who have the ability, resources and inclination to oversee a major project will, or should, use this procurement system unless they are prepared to appoint a consultant project manager to protect their interests.

The 'Report and guidance' document issued by the Construction Management Forum [5] identifies five stages within the construction-management process:

1 concept;
2 detailed feasibility;
3 scheme design;

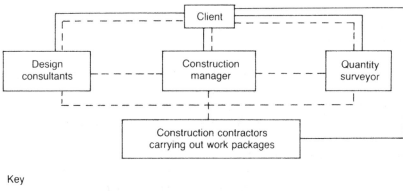

Key

——————— Contractual relationship
– – – – – Functional relationship

Figure 6.4 Functional and contractual relationships, construction management

4   design completion and construction;
5   completion.

These are now discussed in outline.

Concept

During this stage, an initial project brief, including cost, quality/functionality and time parameters, is prepared, sketch designs are produced and a feasibility study is prepared by the designer and construction manager, who, under ideal circumstances, should be appointed during this phase. Such appointments can be limited in time and scope in order to protect the client against the consequences of the project not proving to be feasible.

Elaborate procedures were developed for the selection of the designer and construction manager during the initial use of the system in the UK, with their unusual thoroughness probably reflecting the comparative novelty of the use of the approach at that time. Simplified methods such as those contained within the Construction Management Forum's guidance documents [5] are now available.

Current procedures are similar to those used for the selection of management contractors, with the eventually short-listed organisations making formal submissions, consisting of an offer to provide a specified service for a specific fee, on the basis of more detailed information about the project provided by the client.

Once formal submissions are received, each firm is interviewed in order to discuss the submission and allow the client to meet the designer/construction manager's key personnel. The normal form of interview is a presentation by the firm followed by questioning by the client in order to identify weaknesses, obtain missing information and evaluate the organisation's personnel and capabilities.

Finally, the choice of the designer/construction manager is made by the client based on a full discussion and evaluation with his/her advisers of the short-listed organisations' strengths and weaknesses against a list of weighted criteria.

The designer's duties will include all of those responsibilities normally associated with building projects, i.e. the preparation of all stages of the design, obtaining approvals from statutory authorities and other third parties, preparing cost plans, etc. In addition, he/she will be responsible for co-ordinating the work of other consultants and advising on their appointment, checking and approving any design work carried out by the works contractors, contributing to the preparation of the project brief and generally working closely with the client and construction manager on all aspects of the project.

None of the professional institutions have, up to the present time, produced a model form of agreement for the appointment of the designer when engaged upon a construction-management project which has been accepted by clients. It is therefore necessary for one of the parties to produce a bespoke contract or for the existing model forms produced by the various institutions to be amended.

Once appointed, the construction manager's responsibilities, all of which will need to some extent to be carried out in co-operation with the client and the designer, will generally consist of:

1   The overall planning and management of the total project from inception to completion.
2   The assessment of the design to ensure buildability, compliance with budget costs and programme, value for money and economy.
3   The provision of advice on real construction costs and the establishment of cost budgets for the project.
4   Identification of all statutory requirements, advising the client of his/her time and cost implications and suggesting alternative methods, if necessary, of meeting these requirements and including the establishment of a safety policy for the site.
5   The planning, management and execution of the construction phase of the project, including dividing the work into appropriate packages, obtaining tenders for each works package and ensuring that preordering of essential materials or equipment is undertaken.
6   During the currency of the project, ensuring that the client is kept informed on progress, costs and levels of future expenditure; evaluating and issuing variations once they have received the client's approval; agreeing interim payments and final accounts with the works contractors; and advising the client on contractual claims.

These duties must be incorporated into some form of legal agreement which will establish a contract between the client and the construction manager and determine the roles, responsibilities and liabilities of each of the parties.

In the UK, standard forms of contract do not exist for use for the appointment of construction managers, and specifically designed contracts are therefore usually produced on a project-by-project basis and tend to include very detailed schedules of the construction manager's responsibilities.

It has been suggested that the method of reimbursement of construction managers for their services is somewhat confused as a result of their responsibilities varying from project to project and by the common practice of dividing the remuneration into two separate parts – the first a percentage fee for management services and the second normally a lump sum for the provision of general site facilities and site-based management. It is maintained that this method of payment mixes together consultant and contractor responsibilities and thus makes it difficult to separate management costs and undermines the creditability of the construction manager's consultant status.

This concern can be partially mitigated by the level and fixed cost of the construction manager's site-based staff being agreed in advance and the provision of common services and facilities being made the subject of a works package. The fee is normally fixed and predetermined at the outset and covers the construction manager's overheads and profit.

Detailed feasibility

During this stage, the designer develops the sketch designs and, together with the client and construction manager, enlarges and firms up the project brief, the cost plan and the means by which the project will managed and controlled in order to determine finally whether or not the project is still viable. As a result of this exercise, the client decides whether the project should proceed to the next stage.

Scheme design

The scheme design is finalised during this stage once it has been determined that it satisfies the project brief and can be implemented within the cost plan. If this is the case, approval to proceed is given by the client.

The construction manager will primarily be involved in the preparation of the project cost plan and with forecasting the client's cash-flow requirements. He/she will also be defining the works packages, preparing tender documents, selecting potential works contractors, establishing the project programme and construction method statement and finalising management procedures.

This is the critical stage in terms of determining whether the client is in a position to authorise the commencement of completion of the design and in terms of obtaining of the first bids to allow work to commence on site. Such authorisation can only be given once assurances are received from the designer and construction manager that the client's brief has, and will be, fulfilled.

Design completion and construction

During the early part of this stage, the design is completed and the construction manager begins to implement the tendering procedures by obtaining, reviewing and evaluating bids from works contractors and, together with the designer, making recommendations to the client for acceptance of tenders.

Construction managers do not carry out any of the construction work themselves, but they are responsible for the control and co-ordination of package contractors who actually implement the works under a series of direct works-package contracts with the client. The construction manager acts under this arrangement as the agent of the employer.

Works contracts tend to be private forms based upon amended standard forms of contract, although there is a growing use of the 'JCT with Contractors Design' form as the basis for contracts between clients and works contractors, particularly where the latter has design responsibility.

During the construction period, the main tasks of the construction manager are the controlling of the cost of the project against the agreed budget, estimating the cost of design and construction proposals, using value-engineering techniques to review design proposals, monitoring tender costs and adjusting the content of future works packages to ensure adherence to the approved estimate of the cost of the work.

Time management, or programme control, is another fundamental task for which the construction manager has responsibility – its purpose, having established a programme for the project, is to ensure that the work is completed within the time agreed with the client by monitoring progress of the individual works packages and correcting any deviations from the established programme.

Apart from completing the design, and dealing with the construction manager with any necessary variations to the original scheme, the designer's main task is to examine and co-ordinate the designs submitted by other consultants and the works contractors, ensuring that they comply with the original specifications and the project brief and, if satisfactory, sanctioning their implementation.

As the designer is not always the best person to ensure the quality of the work on site, the construction manager should take responsibility for overall quality control. The whole project team must be involved in establishing the initial standards for works contractors and subsequently monitoring quality and ensuring defective work is remedied at the expense of the perpetrator.

Interim applications for payment submitted by the works contractors are examined and assessed by the construction manager, together with the designer and other appropriate members of the design team, and recommendations made to the client for payment.

The validity on any claims for variations, delays, etc. submitted by the works contractors is established by the construction manager, who assesses and agrees their value and recommends the necessary action to the client after discussion with the appropriate designer. Final accounts are dealt with in the conventional manner.

Completion

The final elements of the construction process, i.e. the inspection, testing, commissioning and handover of the separate works packages once substantial completion has been effected, the issue of individual completion certificates to works contractors and the monitoring of defects liability periods are the responsibility of the construction manager working in unison with the other members of the professional team.

The product

The three criteria of functionality/quality, time and cost are now examined in the context of the performance of the construction-management procurement system.

Functionality/quality

There is no evidence to support the view that there is an improvement in the quality of the project's detailed design as a result of the involvement of the construction manager during the initial stages of the development. Indeed, it has been suggested that contrary to this widely held belief the construction manager's staff often, if contractor trained, has a poor understanding of the design process and consequent commercial bias, and is often unable to adopt an objective approach to design matters.

On the question of the manager's contribution to the 'buildability' of the project, Bennett [17] found when examining a number of projects during a 1986 study that 'the detailed designs observed were no easier to construct than those produced in conventional practice'.

However, provided that the construction manager implements the correct quality-control procedures during the construction stage to ensure that works contractors are aware of the standards expected of them, and that their performance is strictly monitored, a high level of quality can be achieved and, if the manufacturing industry experience is repeated, will be accompanied by an increase in productivity.

Overall, therefore, the combination of an experienced design-orientated construction manager and the correct construction quality procedures provides a good chance of ensuring better-than-average performance in this area of the project's implementation. This view appears to be supported by the Construction Round Table's guide to the building process [16], which suggests that a 'prestige' level of quality can be achieved when using this procurement system.

Time

The construction-management method has two basic characteristics which should lead to faster completion.

First, construction is undertaken by separate works contractors with each works package being capable of beginning as soon as firm designs are available for that package, and the successful tenderer appointed. Thus, project design and construction can be overlapped. This approach can be reinforced by clients preordering and procuring critical materials and components ahead of construction and the fact that construction managers who are normally responsible for the time management, programming and monitoring of the design activities can ensure that information delays are minimised.

Second, the use of this system should ensure that designs are more easily built as a result of the construction manager's input at design stage and the consequent avoidance of unnecessarily complicated work and reduction of overlapping between individual works contractor's operations. The reader will appreciate from previous comments regarding the efficacy of such involvement by management contractors that this advantage can sometimes be more imaginary than real and is only successful when the manager's staff is expert and experienced in this field.

A number of government-sponsored and commercially sponsored reports, when comparing conventional methods of procurement with unconventional approaches such as construction management, confirmed that, on average, the design and construction phases of projects using the unconventional route were completed more quickly than when using conventional procurement systems, and *Thinking About Building* [16] advises the use of this method when early completion of the project is important or crucial.

An examination of contemporary reports on construction-management projects reveals that those involved with the project believe that the enlightened management approach being adopted on site and the fact that a positive attitude to difficulties and problems has been engendered has led to the dramatic improvement in speed of construction over conventionally managed projects.

It has been suggested that this discriminating approach stems from the initial involvement of American clients and construction managers, whose very positive approach has persuaded British consultants and contractors to adopt a more constructive attitude to the management of projects and the solving of specific problems.

Cost

The use of works packages ensures that competition can be achieved on a major part of any project being carried out using construction management. The fact that the employer enters into direct contracts with the works-package contractors ensures a greater measure of control over costs and the overall financial state of the project. It is also possible, during the construction phase of the project, to adjust the scope or specification, and thus the cost, of the uncommitted work should the contract already awarded have exceeded its estimated cost.

Research [18] has shown that the construction-management projects examined had a lower average cost than the management contracts that were surveyed, and that the average cost overrun for the former method was lower than the corresponding figure for the latter. However, very little further specific evidence appears to exist with regard to the comparative cost of this method, and it must therefore be a matter of conjecture as to its relationship with other procurement methods in this respect.

The use of value engineering by construction managers has been cited as a means of improving the value for money provided to clients when using this system. It has been suggested in literature by some advocates of this technique that capital cost savings of between 5 per cent and 10 per cent have been achieved in return for additional fees of 1 per cent with no reduction or, in some cases, even an increase in value.

Although the question of the level of direct construction costs remains unresolved, there is a real possibility that the effect of the likely saving in time that will be achieved when using this method will mean that in terms of the client's total real costs a net saving may be achieved as a result of reduced financing changes, earlier rental income, etc.

The system's association with the fast-track approach, which is demanding of management skills and can lead to the need for heavy expenditure on resources in order to achieve the necessary extra speed of construction, makes for extreme difficulty in reducing direct costs when using this method. In practice, such cost benefits that can be achieved depend upon more mundane day-to-day management activities such as the close scrutiny of the works contractor's accounts (so as to ensure that overcharging does not occur because of complex internal accounting procedures), contra-charges or any other aspects of normal commercial contracting.

One of construction management's main weaknesses is that it leaves the client vulnerable when, at the commencement of the project, he/she has entered into the first works contract and is irrevocably committed, but in return has no guarantee as to what the final cost of the project will be. This shortcoming has to be accepted by the client if he/she wishes to maintain the system's philosophy of the construction manager's responsibilities being limited to those of professional negligence.

There is, of course, no lack of evidence that on many projects various forms of incentives, designed to increase the construction manager's risk, have been implemented in an effort to overcome this weakness and to ensure that the estimated final cost is not exceeded. In these cases, it is likely that the careful detailed planning and costing of the project, which is normally characteristic of this method, will be carried out with even more caution, with the result that the usually high preconstruction costs associated with this method of procurement will be further increased.

There is thus very little evidence to suggest that the use of construction management will result in large cost savings, or the converse, and it therefore appears that the best that can be said is that this system, when correctly used, ought not to perform worse than conventional methods and on large, complex and innovative projects is likely to produce better results in terms of cost than when using many other procurement methods.

General

The emphasis on management which is inherent in this approach generally results in clients and designers making timely decisions to match the needs of construction and, should the need for design changes occur, being in a position where a concerted effort by the whole project team can be made to minimise the time and cost penalties that could be incurred by the client.

As in the case of proposed late changes in the design, many of the benefits arising from the use of construction management centre on the project team's knowledge and experience working unambiguously in the client's interest throughout the life of the project and leading to improved buildability, the minimising of problems during the construction period and the close matching of the contents of the works packages to the capabilities of the works contractors.

The method also provides for strict control of the design process, particularly in terms of the co-ordination of the production of design information with the requirements of the construction programme. Anecdotal evidence also indicates that a good sense of team work and a positive approach to problem-solving is engendered when construction management is used.

The last characteristic may be the result of the greater client involvement that the use of this system demands and the better working relationships that result from this, including those between the construction manager and the works contractors that stem from the client entering into direct contracts with each individual contracting organisation.

The fact that under this arrangement the client is responsible for direct payment of the works contractor's accounts usually results in an improvement in the cash-flow position of the individual contractors as they are paid earlier than when operating under other methods of procurement, on a 'pay when paid' basis, when main or management contractors have this responsibility.

Clients should be aware, however, that the use of construction management involves them in additional administrative duties and responsibilities to those accepted when using most other methods of procurement and also increases the risks they carry, particularly those associated with cost overruns, delays and claims.

The fact that they are often faced with the need to make timely decisions in their more involved role as the leading member of the project team may be unattractive to those clients used to the more leisurely style of consultation with their professional adviser which characterises conventional practice.

Despite their self-proclaimed expertise in the area of buildability, contractor-based construction managers are often unable, because of lack of experience on the part of their staff in this field of activity, to contribute in a constructive and sensitive manner to the improvement of the design of the project.

Despite the advantages gained by works contractors when involved in construction-management projects, it is possible that they can feel more exposed when tied to a direct contract with the client and lack the protection of a main contractor. More of the responsibility for co-ordination and performance rests with them, as does much of the site management and 'attendance' tasks previously undertaken by the main contractor on more conventionally managed projects, and under certain circumstances they can also be allocated onerous design responsibilities.

As is the case with all building procurement systems, the circumstances of each project and the competence and experience of the design and construction teams need to be weighed against the advantages and disadvantages that have been identified as characteristic of construction management before a decision can be made as to their suitability for use on a specific project.

However, this method needs to be executed with great care and precision in order to ensure that the potential benefits, particularly speed of construction, can be achieved. On the basis of its track record, construction management if used correctly and efficiently has become firmly established as one of the most efficient methods of successfully managing large, complex building projects.

In this context, the report published by the Centre for Strategic Studies in Construction [5] concluded that the aims of a client who wishes to follow the 'management path of procurement' are best achieved through the construction-management method.

6.4 Design and manage

Definition

In this procurement system, a single organisation is appointed to both design the project and manage the construction operations using package contractors to carry out the actual work.

The single organisation responsible for the project can be either a contracting organisation or a consultancy practice, with the former being elevated to the status of a consultant upon appointment, although the client may well employ an in-house or consultant project manager to supervise the project and will invariably appoint an independent quantity surveyor to oversee the financial aspects of the scheme. The main common characteristics of the system are that:

1   A single organisation is appointed to both design and manage the project.
2   The single organisation can either be a contractor or a consultant.
3   The actual construction of the project is carried out by works, or package, contractors who, in the case of the contractor variant, are employed by the contractor; when using the consultant variant, the works contractors are directly employed by the client.

Reimbursement is by means of a lump sum or percentage management fee with the actual cost of the works packages, together with any common services, being paid to the contractor where he/she is responsible for the management of the project.

In this guide, the design and manage system has been thus defined:

> The design and manage organisation acts as a consultant, normally with direct responsibility to the client for the design and construction of the project.
>
> All construction is carried out by means of works packages, which are either the subject of direct contracts between the client and the package contractors (consultant variant), or contracts between the design and manage organisation and the package contractor (contractor variant).
>
> In the former variant, reimbursement is by means of a professional fee and in the latter by means of a fee together with the actual cost of common services and works packages.

Figure 6.5 illustrates the alternative functional and contractual relationships observed in this procurement system.

Contractor-led design and manage

When using this variant of the system, clients usually obtain single, or sometimes two-stage, competitive bids and proposals from a small number (generally three or four) of suitable contractors on the basis of an enquiry document containing details of the client's requirements, conceptual designs, etc.

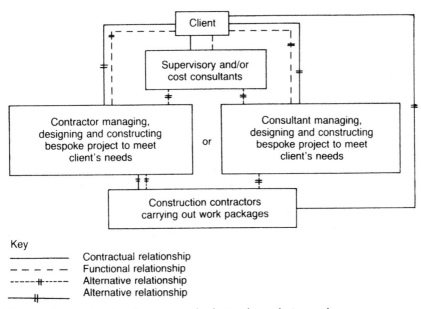

Figure 6.5 Functional and contractual relationships, design and manage

Tenderers are usually required to state their lump sum design and management fee and offer a guaranteed maximum price (GMP) with their submissions, and the contract award is normally made on the basis of the financial level of the bid and the tenderer's design proposals.

The requirement for tenderers to be tied to a GMP reflects the approach sometimes adopted by clients when using management contracting and which has already been discussed, and deprecated. When the client wishes to receive and benefit from independent management advice from a contracting organisation adopting a consultant role, there is an essential need to divorce the contractor from the profit motive associated with the conventional approach to construction projects. The total loyalty to the client's objectives essential to the success of management-orientated procurement systems is undermined by the insistence on the GMP requirement being incorporated into any agreement between employer and contractor and as such should not be demanded of any contractor when using the design and manage system.

Design of the project can, as in the design and build method, be carried out by in-house designers, external consultants or a combination of both, but whichever type of resource is used the responsibility for the total design rests firmly and comprehensively with the design and manage contractor.

Close co-operation between the client, the design and manage organisation and, in some instances, works-package contractors on matters of detailed design is a characteristic of this method, with all the main parties – particularly at crucial states of the project – often locating technicians and designers on site so as to resolve problems that may arise and to carry out the detailed design of the scheme.

All of the construction work is carried out by means of trade, or sectional, works packages, which enables design and construction to be overlapped with all of the benefits that such an arrangement brings. Provision is sometimes made for the design and manage contractor to provide site-establishment works and common-user services from his/her own resources; this approach should be discouraged, and the necessary work implemented by means of packages, in order once again to ensure that the contractor's consultant status is not undermined.

The best-known example of the use of this variant of the method is probably the design and construction of the £25 million Nissan car plant complex at Washington, Co. Durham, carried out some 15 years ago when the following note [19] was circulated to tenderers:

> Whilst the system is intended to encourage early resolution of every aspect of design, the progressive letting of works packages gives clear opportunities for the design of services to be delayed until a fairly late stage. It is intended that this opportunity should be used so that client/contractor dialogue can continue as long as possible in the interest of obtaining services precisely matching the client's present, or future, needs.

While this approach illustrates one of the main advantages gained when using this system, it also highlights the difficulties that can be caused by the ability of all the management-orientated methods to cope with the client's desire to postpone decisions until the last possible moment. Apart from the difficulties of providing an accurate forecast of the final cost of the project and the effect that any late changes are likely to have on completed or current work, there is always the real likelihood that delays to the overall progress of the project can occur, and clients need to be made aware in the strongest terms of the dangers inherent in implementing late amendments to the original design. Latest dates must be agreed by which time such alterations, if essential, can be allowed.

In the majority of projects where this variant of the system has been used, the client has appointed an independent financial consultant, usually a professional quantity-surveying practice, to provide advice on the financial and contractual aspects of the scheme. Once the design and manage contractor is appointed, the formulation of works packages, obtaining of tenders and financial monitoring and control of the project is the subject of joint management and liaison between the two parties, although the contractor remains contractually responsible for all aspects of package contractor management.

The design and manage contractor is usually reimbursed the sum of:

1   the design fee;
2   the management fee;
3   the fee for, or the net cost of, site establishment and common user services;
4   the net total of all the works contractor's final accounts up to the GMP, if appropriate.

Where a GMP is incorporated within the contract, it is usually subject to adjustment in accordance with the precontract agreement, which will have been described within the tender documentation. In essence, such an agreement will reward the contractor for any savings made against the GMP and penalises him when this sum is exceeded as a result of his/her own mismanagement or negligence.

Payment is generally made on the basis of the net cost of works contractors accounts by means of the 'open book' method of accounting, whereby the design and manage contractor's financial records of the project are made available to the client, or his/her financial advisor, in order that the actual cost of the works can be verified and interim certificates issued in the conventional manner subject to the standard retention rules which are usually applied.

The question of quality control is a vexed one, with the responsibility for this function nominally being allocated to the design and manage contractor but in reality usually devolving upon quality controllers, generally in the form of clerks of works or inspectors, independently appointed by the client.

This scenario appears to be the norm for those projects where the consultant contractor's status is undermined by the client's desire to try to obtain the best of both worlds by introducing a commercial element into the contract as an incentive to the contractor to minimise costs while at the same time requiring him to act as an independent consultant. Under these circumstances, there is always the possibility that quality may well suffer unless additional supervision is brought in to monitor the standard of workmanship and adherence to specification of materials and equipment.

Apart from the costs incurred in employing and managing the supervisory staff, this approach must undermine the close relationship between the client and the design and manage contractor that is intended to form the fundamental basis for the success of any project, where this particular procurement system is being used.

The literature shows that a number of subvariants of the contractor-led system have, and are being, used by clients. The most common of these appears to be the design, manage and construct approach [6], in which the contractor is divorced in part from his/her role as a consultant to the client and allowed to take on some of the construction work, using his/her own or subcontracted labour, and accepting contractual responsibility for meeting project time, cost and quality targets.

Consultant-led design and manage

This variant is also referred to, in literature, as the alternative method of management (AMM). The fundamental difference between this and the contractor-led variant of the system is that, in this case, it is the client who enters into a contract with each individual works-package contractor, leaving the consultant with the comparatively risk-free responsibility for the design and management of the project.

While contractor-led design and manage reflects a practical – and often highly commercial – approach to project implementation, AMM was originally conceived as the application of a philosophy, and a set of principles, to the construction process. The philosophy mirrors the system's creator's belief that the efficiency of most construction projects suffers as a result of poor communications, an unnecessarily large and complex management hierarchy, overplanning of the project and lack of commitment to the client's cause by most of the participants. The principles for the management of a project that were established can therefore be summarised as follows.

1   Documentation, particularly drawings, are not a product but a means of efficiently communicating the client's requirements to the implementer.
2   Communications among client, consultant and works contractor/ tradesmen should be as direct as is practically possible.
3   As changes are inevitable on construction projects, full preplanning is not the best means of control and a more flexible and less-detailed short-term approach to programming should be adopted.

4   The client's interest is best served by all of the participants being committed to the project objectives not their own individual aspirations.

However, in replacing the contractor on site, the consultant is placed in an unfamiliar role and, apart from a high level of competence, must possess exceptional personal qualities, social skills and substantial and continuing support from his/her practice office. In the case of consultants who are not operating out of a multidisciplinary practice, it will be necessary for the missing expertise to be bought in from an external source. In most of the recorded cases, this buying in has tended to be for the quantity-surveying discipline.

As there are a limited number of professional practices that have the necessary attributes and experience to design and manage complex construction projects successfully, clients need to make certain that short-listed consultants are meticulously examined so as to ensure they are capable of carrying out a function which demands talents that are not necessary when acting as designer, financial adviser, design-team leader or even the client's project manager.

The majority of consultant practices will carry out both traditional design work and design and manage projects, and the possibility of conventional design attitudes influencing staff working on this unconventional approach to managing projects is always present. The accepted practice of building a 'Chinese wall' – i.e. an imaginary division restricting communication between two activities housed in the same building or organisation – may be effective, but clients need to satisfy themselves that the attitudes of the staff that will be involved in their project are appropriate for the procurement system being used.

As the client is responsible for the direct reimbursement of the various works-package contractors, the only payment made to the design and manage consultant is for the cost of the agreed fee for designing and managing the project. The site establishment and common-services elements are carried out by works contractors as one or more of the packages.

Some consultants are prepared to provide the client with a guarantee that the project will be completed on time, and within the estimated cost, and to accept penalties and incentives for failure or success in achieving the agreed objectives. Once again, the introduction of commercialism by the adoption of this approach could be seen as undermining the credibility of the consultant as an independent, objective adviser to the client.

According to the limited amount of literature available, clients appear less prone, when using the consultant variant, to appoint independent quality controllers than when adopting the contractor-led alternative, and where the consultant is not tied to any penalty, or bonus, this approach may be valid, particularly when the consultant is experienced in running design and manage projects. However, clients should make certain that the quality-control aspects of the project are examined before work commences and that a comprehensive monitoring regime is established.

The ability of this system, in common with others in the management-orientated category, to enable work to commence on site before the total design has been completed has been identified as the main advantage offered by this procurement method.

Thinking About Building [16] confirms that when early completion of the project is crucial both forms of the system will satisfy this requirement. CIRIA's *Planning to Build?* [15] refers to the benefits of overlapping of activities and states that project implementation can be quick when using this method.

In the case of the Nissan project, the contractor-led team completed the design and construction of the complex scheme in a total time of 18 months. If conventional methods of procurement had been used, the client's quantity surveyor expressed the opinion that this period would have been appreciably longer and the task of implementation more difficult and less efficient.

A further common feature of both variants is the presence on site, generally on a semipermanent basis, of the personnel responsible for the design of the project, whose duties while resident include further detailed design, clarification of existing design details, liaison with works contractors to ensure buildability and working with the client's representatives to make certain that the project's functional requirements are maintained and achieved.

The shortened lines of communication between the various parties, closer understanding of the design by the works contractors and rapid decision-making engendered by this characteristic of the system is beneficial particularly in the improvement that is achieved in the relationship between the client and the design manager and that between the latter and the works contractors.

It should be noted, however, that the presence on site of the designer, which is essential for maximum success, can be both time-consuming and expensive and requires a total commitment from the designer manager, which is not always achieved.

In the same context, the need for close client involvement with the project, particularly with regard to design matters, and the requirement that decisions may often have to be made very quickly can apply considerable pressure to some clients and/or their representatives with the result that the relationship between the employer and the design and manage organisation, which should have been enhanced by the use of this particular system, can become strained.

The fragmented and overlapping nature of this procurement system allows any changes introduced by the client to uncommenced works packages during the construction period to be incorporated provided they are implemented before tenders for the appropriate package are obtained. Any attempt to introduce such amendments after tenders have been received, or work commenced, must normally result in delays and additional costs being incurred.

Cost monitoring and control can be more strictly implemented when using the design and manage system than when other procurement approaches are adopted, mainly owing to the ability to redesign and continue to make cost reductions for the remaining elements of the project. The overlapping of design and construction inherent in this method allows the design to be amended to reduce the cost of a particular works package if previous costs have proved more expensive than envisaged.

When using this method, it is necessary that the client be prepared, as with the other management-orientated procurement systems, to commit himself/herself to the project without any guarantee (unless the GMP approach is adopted) of the final financial outcome. It is therefore likely that the private sector will appreciate the value of the savings in time that can accrue, and will be prepared to accept the need for this commitment, under conditions which would not normally be acceptable to the guardians of public accountability.

There are no standard forms of contract in existence to formalise the legal relationship and responsibilities between the two main parties and little evidence exists as to the actual documentation used on completed projects, but it can be fairly safely assumed that in the case of the contractor variant the contractor's own form is used. The Association of Consultant Architects (ACA) contract forms have been designed to be used with the consultant-led alternative, and this document and various JCT forms, often heavily amended, have been utilised for the works-package element of the project.

Detailed conclusions with regard to the use of this procurement system will be reached at a later stage, but it should be said in the meantime that this method, when properly implemented, is capable of being used successfully on a wide range of types, sizes and values of projects and could well be more widely used by clients to their advantage.

6.5 Summary

Common characteristics

The common characteristics of all three of the management-orientated systems previously described are now identified.

Advantages

1   The use of this category of systems enables commencement of the project to be accelerated, which, in turn, should enable earlier completion to be achieved than when using procurement systems in the separated categories.
2   Early advice can be obtained from the contractor/manager on design, buildability, programming and materials availability, together with general construction expertise.

3   The systems within this category have a high degree of flexibility to allow for delays, variations and rescheduling of works packages.

4   Because, when using the systems within this category, the financial structure of the project is fragmented, the monetary failure of any works contractor will only have a limited effect on the total process.

5   The use of individual works packages to carry out all construction work ensures that competition can be achieved on up to as much as 90 per cent of the construction cost of the project and makes it possible to adjust the cost, or scope, of uncommitted work should the packages already awarded have exceeded their estimated cost.

Disadvantages

1   One of the fundamental aims of this category of procurement system is the elevation of the contractor to the status of client's adviser/consultant, with the result that the contractor's contractual liabilities are limited, in the same way as other members of the professional team, to accepting responsibility for any negligence in the performance of his/her management function.

   All of the systems within this category allocate the majority of the project's risks to the client. These can be particularly onerous where works-package contractors fail to perform and affect following and parallel operations.

2   Although the contractor/manager is responsible for supervising construction and ensuring that work is built to the standards identified by the design team, the fact that his/her obligations are limited to his/her management performance means that the client is liable for the cost of remedying any defects resulting from the substandard performance of any works contractor who is unwilling, or unable, to rectify his/her own faults.

3   The whole question of maintaining quality control is problematic when using the procurement systems within this category, and the client may therefore need to appoint additional site supervision to avoid difficulties in determining the responsibility for defects and to ensure that the specified quality standard is achieved.

4   The client does not have a firm price tender available before commencing work, although both private and public accountability can be partially satisfied as the majority of the construction cost can be subject to competitive tender.

Characteristics of individual management systems

The main advantages and disadvantages of the individual procurement systems contained within this category are now identified.

Management contracting

Advantages

1   A variant of the pure system enables the client to obtain, from the management contractor, a GMP for the construction element of the project.
2   This method is perceived as allowing accelerated project progress and completion, but the results of research are somewhat ambiguous in this respect and it can only be concluded that the risk of delay is reduced and that time targets are unlikely to overrun.

Disadvantages

1   Where a GMP is obtained, the management contractor's status as the client's adviser/consultant is jeopardised and, therefore, there can be the real possibility of a conflict of loyalty.
2   Current research indicates that the project costs incurred when using this system are higher than those generated when using the conventional or design and build systems. This situation results in the main from high tenders submitted by works contractors, as a result of onerous contract conditions and high levels of risk imposed by management contractors, and the duplication of management and common-services costs brought about by the presence of both the management contractor's and package contractor's site organisations.

Construction management

Advantages

1   This method, if correctly applied, can result in a more constructive and positive attitude being exhibited at management, supervisory and operative levels.
2   The fact that the client enters into direct contracts with individual works-package contractors enables a high level of immediate cost control to be achieved and also ensures that the works contractors' cash flows are improved as a result of receiving direct payments from the client rather than through an intermediary.
3   The increasing use of value engineering by construction managers is seen as a positive means of improving the value for money provided to clients.
4   The client's increased involvement in the management of the project, when compared with other methods, promotes better working relationships within the project team.

Disadvantages

1   The nature of this system requires the client to be deeply involved at all stages of the project, to have sufficient in-house expertise to be able to co-ordinate the activities of the construction manager and the design consultants and to be able to accommodate the additional administrative duties and responsibilities inherent in the system.
2   The present position on fees is confused as the construction manager's responsibilities vary from project to project and the division of reimbursement into a percentage fee and a lump sum for the provision of common site services can jeopardise the construction manager's status as the client's consultant and lead to a conflict of loyalties.

Design and manage

The single organisation responsible for the design and management of the project when using this system can be either a contractor or a consultant. The common characteristics of the two methods are now described, followed by the individual characteristics of each system.

Common characteristics

Advantage

The on-site presence of the project designers for a substantial part of the construction period is beneficial to the project in that it results in shortened lines of communication, closer understanding of the design and rapid decision-making.

Disadvantage

The near permanent presence on site of the designer(s), which is essential for maximum success, can prove costly and time-consuming and requires an above average commitment from the designer/manager.

Contractor-led design and manage

Advantage

When using this variant, the client can gain the advantage of obtaining a GMP for the construction element and a fixed price lump sum for the design and management of the project.

Disadvantage

When taking advantage of the contractor's GMP offer, the client may well jeopardise the contractor's status as the client's consultant with the result that a conflict of loyalties may occur.

Consultant-led design and manage

Advantage

Many consultants are willing to agree a fee structure with the client which will incorporate cost penalties and incentives, which can be applied if agreed criteria are, or are not, achieved.

Disadvantage

When acting in a lead capacity, and in essence replacing the main contractor, the consultant is placed in an unfamiliar role which requires a high level of managerial competence and experience, and many designers are unable to meet the demands of this unusual role.

The way in which these characteristics are taken into account when selecting procurement systems is dealt with in Chapter 9. Suffice it to say, at this stage, that the systems contained within the management-orientated category are likely to be most appropriate on large, complex, fast-moving projects where the client has sufficient internal resources to enable him/her to become involved in the management of the project.

References

1   Sidwell, A.C. (1982) 'Management contracting', paper presented to a seminar on management contracting, Institute of Quantity Surveyors, 1 December, London.
2   Heery, G.T. and Davis, E.M. (1976) 'Construction programme management', *Building Technology and Management*, December, 22–26.
3   Flanagan, E. (1987) 'Procurement attitudes in the nineties', paper presented to a seminar, Industrial Development Agencies, 13 February, Glasgow.
4   Carter, J. (1973) 'Management contracting: the Horizon Project', *Architects Journal*, 14 February, 395–400.
5   Construction Management Forum (1991) 'Report and guidance', Centre for Strategic Studies in Construction, University of Reading.
6   Curtis, B., Ward, S. and Chapman, C. (1991) *Roles, Responsibilities and Risks in Management Contracting*, Construction Industry Research and Information Association special publication no. 81, London: Construction Industry Research and Information Association.
7   Naoum, S.F. (1988) 'An investigation into the performance of management contracting and the traditional methods of building procurement', unpublished PhD thesis, Brunel University.
8   Davis, Langdon and Everest (2000) *Contracts in Use, a Survey of Building Contracts in Use in 1998*, London: Royal Institution of Chartered Surveyors.
9   Dearle and Henderson (1988) *Management Contracting – a Practice Manual*, London: E & FN Spon.

10 Joint Contracts Tribunal (1987) *Practice Note MC/1, Management Contracts Under JCT Documentation*, under review, London: Building Employer's Confederation.

11 Sidwell, A.C. (1983) 'An evaluation of management contracting', *Construction Management and Economics*, 1, 45–46.

12 Barnes, M. (1984) *Management Contracting for Health Building – a Comparative Study*, London: Department of Health and Social Security.

13 Centre for Construction Market Information (1985) *Survey of Management Contracting*, London: Centre for Construction Market Information.

14 Naoum, S.F. and Langford, D. (1987) 'Management contracting – the client's view', *Journal of Construction Engineering and Management*, September, III [3].

15 Construction Research and Information Association (1995) *Planning to Build? A Practical Introduction to the Construction Process*, Construction Industry Research and Information Association special publication no. 113, London: Construction Industry Research and Information Association.

16 Construction Round Table (1995) *Thinking About Building*, London: The Business Round Table.

17 Bennett, J. (1986) 'Construction management and the chartered quantity surveyor', *Journal of the Royal Institution of Chartered Surveyors*, March.

18 Olashore, O.B. (1986) 'Management contracting and construction management – a comparative analysis', unpublished MSc dissertation, Brunel University.

19 Lush, E. (1986) 'Fast tracking for a Japanese company', paper presented to the Project Procurement and Management seminar, July, Birmingham.

7 Discretionary procurement systems

7.1 Introduction

Until now the categories of procurement system that have been proposed have been determined by the way in which the design, construction, finance and other elements of the project are related and/or managed.

However, the emergence of a technique for managing the cultural environment of projects called *partnering* together with the introduction in 1983 of the *British Property Federation (BPF)* system, neither of which fit into any of the existing categories, suggests that a fourth classification should be established.

Within this latest suggested grouping, the client has the discretion to use any of the systems from any of the other categories either singly or in combination, or even a bespoke system of his/her own making but with the chosen system(s) being implemented within a specific setting controlled by the client.

A discretionary system is, therefore, an administrative and cultural framework into which any procurement system(s) can be incorporated, thus allowing the client to carry out the project by imposing a very specific management style, or company culture, while at the same time enabling him/her to use the most suitable of all of the available procurement methods.

It could be argued that any method included within this category is not, in reality, a procurement system but rather a means of controlling the project environment. However, the definition established by CIB W92 states that such systems are '... a strategy to satisfy client's development and/or operational needs...' and both of the procurement paths included so far within this category fall within that definition.

Both the use of partnering in the UK and the formulation of the BPF system were instigated by client organisations in an effort, by changing their approach to the management of projects, to overcome what they saw as the unsatisfactory and substantial shortcomings of the contemporary construction industry when implementing projects.

7.2 The British Property Federation system

This method of procurement – which has now been in operation for nearly two decades – appears to have been little used by clients, to the point where the Federation no longer publishes the form of contract designed to accompany the use of the system [R. Kauntze (Deputy Director General British Property Federation), personal communication, 1999].

Notwithstanding this apparent lack of use, a brief examination of the system, the reasons for its birth and the industry's reaction when it was first initiated is considered to be worthwhile.

The system was designed, according to its operating manual to:

> ... change attitudes; produce good buildings more quickly and at a lower cost; create a fully motivated, efficient and co-operative building team; remove overlap of effort... which is prevalent under the existing systems; redefine the risks so that the commercial success of the designer and the contractor depends more on their abilities and performance; re-establish awareness of real costs... and eliminate practices which absorb unnecessary effort and time and obstruct progress towards completion.

The following summary attempts to encapsulate the primary characteristics of the basic system and provides an overview of a relatively complex process:

1   A client's representative is appointed together with a design team leader and any other necessary consultants who design the project in whole, or part, and are reimbursed by means of a lump sum fee together with bonuses if preagreed design production targets are met.
2   Tenders are invited using specifications and drawings, but no bills of quantities. If any elements of the project have not been designed before this stage, they are comprehensively described within the tender documentation in order that tenderers can design the residual sections or elements.
3   Tenders are submitted on a lump sum basis, with a fixed price submission being mandatory for projects of less than 2 years' duration, and a priced schedule of activities with a detailed programme being provided for every activity specifying the resources and construction methods that are proposed. Any necessary design submissions have to accompany the tenders.
4   The contract for the project is normally awarded to the tenderer who submits the lowest acceptable bid.
5   A supervisor is appointed by the client before the works commence on site whose function is to ensure that the project is constructed and, where appropriate, designed in accordance with the contract documents.

6 Payments to the contractor during the currency of the project are made on the basis of the results achieved, i.e. once an operation listed in the contractors' schedule of operations is completed a claim for reimbursement of the value of the work contained within the operation can be made.

7 Disputes between the client and the contractor are immediately referred to an adjudicator for a speedy decision, but if any party disputes this decision it can be referred to arbitration once the project has been satisfactorily completed.

Figure 7.1 shows the contractual and functional relationships between the various parties to this arrangement.

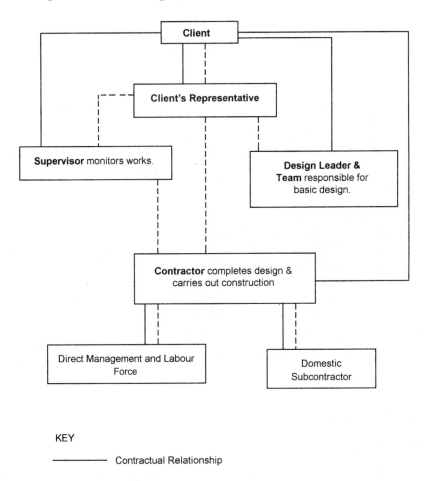

KEY

—————— Contractual Relationship

— — — — — — Functional Relationship

Figure 7.1 The British Property Federation system

The BPF system was conceived in 1980, by which time members of the Federation, an association of property developers, commercial organisations and large retailers with substantial property holdings, had become frustrated and concerned at the frequently occurring problems they experienced when they carried out their building projects.

As a result of this, a working party of senior figures from the Federation representing the most powerful commercial property interests in the UK was established with an urgent remit to '... draft an improved management system for the building process appropriate for members of the BPF' [1].

It is indicative of the approach adopted by the BPF during the gestation period that no formal consultations took place with any of the construction industry's representative bodies, although informal discussions took place with individuals that '... the BPF respected'. This decision was apparently taken to ensure that the new system was innovative and not sullied by the reactionary attitudes of the industry's establishment.

The manual describing the system was eventually published in December 1983 and, during the following months, was received with a mixture of interest, scepticism and some mild hostility. The industry as a whole was curious and sceptical, whereas quantity surveyors individually tended to be more ill-disposed towards the proposals as a result of the use of bills of quantities being actively discouraged by the system. On the other hand, a report [2] produced by the Quantity Surveyors' division of the Royal Institution of Chartered Surveyors during the early part of 1984 adopted a more objective attitude, giving the new system a cautious welcome and suggesting that only use of the system over time would reveal whether it was an improvement on existing procedures.

Doubts were also expressed as to the system's ability to solve the fundamental problems besetting the construction industry and whether certain features of the method would be acceptable, and operate satisfactorily, in practice.

A quarter of a century later, it would appear that the somewhat draconian nature of the system has not become an accepted, or favoured, procurement method, nor has it succeeded in changing the way in which the industry operates. Furthermore, it would appear that other less confrontational approaches may have had more influence for the better on such changes that have taken place.

7.3 Partnering

Introduction

Although the BPF system depended on a somewhat confrontational approach to the management of projects, partnering relies on trust and collaboration and has at its core a philosophy of real co-operation, partnership and equality among all the members of the project team, thus engendering a mutual desire to fulfil the project's objectives.

Partnering is, as has already been established, a means of administering and establishing an environment within which a project is implemented using any of the available procurement systems to carry out the funding, design, construction etc. of the project, although some systems – particularly those from the integrated category – work better with partnering than others.

There are currently two main types of partnering – *project partnering*, where the relationship is put in place on one specific project and terminated once the project is completed, and *strategic partnering*, where a long-term relationship is established which relates to a series of future projects spread over time.

Definition of the system

Partnering has been defined by the Construction Industry Institute (CII) [3] as:

> A long term commitment between two or more organisations for the purpose of achieving specific business objectives by maximising the effectiveness of each participant's resources. This requires changing traditional relationships to a shared culture without regard to organisational boundaries. The relationship is based upon trust, dedication to common goals and an understanding of each other's individual expectations and values. Expected benefits include improved efficiency and cost effectiveness, increasing opportunity for innovation and the continuous improvement of quality, products and services.

The initial reference to a long-term commitment establishes that the definition relates to strategic partnering, but the subsequent recital also adequately and accurately describes project partnering and goes on to list the major benefits that can be expected from the use of both variants of this arrangement.

The Construction Industry Board (CIB) has provided a more succinct definition [4]:

> Partnering is a structured management approach to facilitate teamworking across contractual boundaries. Its fundamental components are **mutual objectives**, agreed **problem resolution** methods and an active search for **continuous measurable improvements**.

The CII's definition identifies three critical requirements for the successful implementation of partnering, i.e. the acceptance of the change from a traditional relationship to that of an unbounded shared culture, the achievement and acceptance of common goals through the establishment of mutual trust and understanding and the distribution of benefits to all of the project participants.

The CIB's definition identifies the three fundamental elements which form the core of the approach:

1   All of the parties involved in the project, or series of projects, including subcontractors and suppliers, work together in partnership to ensure that *mutual project objectives* are met by improving the performance of all of the participants.
2   Methods of *problem resolution* are agreed at the commencement of the project.
3   Every effort is made by all of the participants to ensure that the performance of the project team is *continuously and measurably improved.*

Successful partnering can only be achieved if all of the participating organisations are prepared to trust the other members of the project team and accept that the majority of people are hardworking, respond to a challenge and are basically trustworthy.

Genesis Background

Contrary to current perceptions, partnering, although not necessarily in its present form or using that particular name, has existed within the UK construction industry since at least the early 1900s, when Marks and Spencer and Bovis began a long-standing relationship which has lasted to the present time and is based upon mutual trust and respect as well as the resulting commercial benefits enjoyed by both parties.

This arrangement was one of the first of the many preferred contractor/ supplier arrangements and alliances which have been formed and which have flourished over the years before the demand side of the industry, spurred on by the Latham and Egan reports and concerned at the continuing confrontational attitudes exhibited on construction projects, decided to adopt the partnership approach in an effort to improve the project-implementation process.

In the USA, it was in the late 1980s that the growth of claims and litigation on construction contracts led public agencies to begin to use the technique that led to a promising increase in the controlling of cost and time growth on numerous major projects [5]. Many states and central governmental organisations are now committed to the use of partnering, and it is understood that the growth of the use of the method in the private sector has increased significantly in the recent past and continues to expand.

In Australia, formalised partnering emerged at the beginning of the 1990s, partially, it is believed [6], as a result of central and state governmental initiatives which put forward a new strategy to improve production in the construction industry, which included a commitment to partnering.

The emergence of partnering proper in the UK also appears to have taken place in the early 1990s – the National Economic Development Committee's (NEDC) report *Partnering: Contracting without Conflict* [7] was published in 1991 – although some large client organisations may well have been using the system before then.

Share of the market   *degree of usage*

There is currently very little evidence available as to the level of use of partnering on construction projects in the UK. The latest survey of contracts in use [8] carried out for the Royal Institution of Chartered Surveyors collected information on this system for the first time and found that the method had been used on only forty-two of the total of 2,457 projects (1.7 per cent).

These figures are presumed to relate to specific project partnering arrangements and are probably low, particularly if strategic partnering arrangements are also taken into account. It is also of interest to note that a study by Walter [9] in 1998 confirmed that many of the UK's major construction companies were expecting partnering to account for over two-thirds of their turnover in 1998/1999.

The process

It is important to understand that the use of partnering does not supersede the process used by the procurement system chosen to implement the project, but rather acts as a framework within which the selected system operates more beneficially. The use of partnering is a voluntary arrangement made between all of the project participants, has no legal standing and imposes no contractual obligations upon any of the parties.

While most of the advantages of the method are maximised when using strategic project partnering, the majority of projects upon which it will be used in the UK are likely to be one-off in nature and will thus be implemented using specific project partnering. As the process differs from one to the other, both approaches are now described separately.

Project partnering

The essential stages of the project partnering process are:

1   Decision stage – the desire of all of the main members of the project team to become involved in the partnering process is essential; this desire will often stem from relationships that have been formed during past associations when organisational cultures have been found to be compatible. If this has not been the case, an examination of each of the main participants' organisations and personnel to establish compatibility, or the lack of it, is essential before any commitment is entered into.

Once all of the parties are convinced that they wish to participate in the partnering process, a decision to use partnering on the project under consideration can be made and the next stage of the process can be commenced.

2 Establishment of working practices – during this stage, a series of mutual objectives for the project are identified and agreed and the way in which problems are to be solved during the duration of the project is established.

These two activities are carried out by means of a workshop, which needs to be held immediately the decision has been taken to use the partnering process and should be attended by all of the main parties involved in the project. This event is critical to the success of the project as it will determine how the project team will implement the project together and engender mutual understanding among the participants [10]. This initial workshop is likely to be of 1 or 2 days' duration as it will be dealing with a considerable amount of detail in order to reach agreement on a list of mutual objectives, identify improvement targets, design the problem resolution process and deal with any matters specific to the project.

The final major task of the workshop is to formulate a partnering charter, a document with no legal standing, which incorporates in simple language the agreed mutual objectives, a summary of the problem resolution process and the basic philosophy and aims of the partnership. Once this document has been signed off by all of the parties, it is exhibited in the offices of all the participating organisations and in all site offices, canteens, notice-boards, etc.

At the end of the workshop, arrangements are made for future follow-up and induction workshops.

3 Implementing partnering practices – the execution of the practices agreed during the second stage of the partnering process is carried out in parallel with, but physically separate from, the everyday management of the project by means of the follow-up workshops.

These workshops will take place at intervals and will be of a duration determined by the project team; they will also deal with any problems that have arisen during the implementation of the project, review progress that has been made in achieving the agreed mutual objectives and take whatever action is necessary to ensure that the previously agreed continuous improvements are being made. Subworkshops, sometimes referred to as action teams, are used to deal with matters which are not capable of being dealt with during the comparatively short duration of the normal workshops.

On the appointment of new key members of the project team, whether they be consultants, subcontractors or suppliers, induction workshops are held in order to familiarise the newcomer with project partnering and to obtain their agreement to participating in the process and incorporation into the partnering charter.

Over the often lengthy implementation period of many major projects, these workshops should enable project teams to establish the trust and understanding which is so necessary to improving working relationships and enhancing performance to the benefit of all of the participants.

Figure 7.2 illustrates the project partnering process.

Strategic partnering

Strategic partnering involves the long-term use of partnering by two organisations to carry out more than one construction project or some continuous construction activity such as a maintenance programme. The process is implemented on individual projects as they arise and other organisations are brought into the arrangement on a project-by-project basis [7].

While the strategic partnering philosophy replicates that of project partnering, the fact that the commitment required from both parties is far greater, as a result of the long-term nature of the arrangement, means that more care and time are needed in the initial stage to ensure that both organisations and their respective managements are totally dedicated to the use of strategic partnering.

Although, as a result of its specific nature, the choice of method of procurement and associated contractual arrangements is relatively straightforward when implementing project partnering, this aspect of strategic partnering can often be problematic because of the many ways in which long-term procurement can be carried out and associated contracts awarded. These elements can also be complicated by the need to comply with the requirements of European Union and UK competition law and, particularly in the case of governmental bodies, to satisfy public accountability.

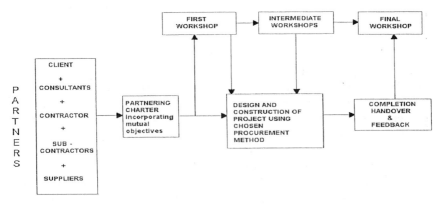

Figure 7.2 The project partnering process

The case studies examined in the report of a research team from the University of Westminster in 1997 [11] illustrate the wide range of procurement and contractual arrangements that can be used in strategic partnering. Procurement methods included conventional, design and build and management contracting; awards were made on a negotiated or competitive tender basis; and, in some cases, a number of contractor partners were appointed to carry out a share of the total long-term workload. In four out of the five cases, standard forms of contract were used, with no contracts being entered into in the fifth case and only one partnering charter being used.

Similar characteristics were exhibited by the partnering arrangements examined in the 1998 University of Reading research study [5], which, having looked at a much larger number of projects, confirmed that there were numerous ways of implementing strategic partnering depending upon the client's culture, circumstances and needs.

So, whereas the strategic partnering process can be said to follow the same path as project partnering through commitment to co-operation by means of workshops, etc., the actual process is likely to vary considerably from case to case, and thus is only capable of being described in the general terms used here.

The product

The performance of project and strategic partnering is jointly examined.

The numerous partnered projects that have been studied in the UK, the USA and Australia suggest that performance indicators in the areas of project cost, speed of implementation, quality, teamwork and communication should be the criteria against which the success of partnering should be judged.

Cost

There is wide variance in the findings of studies into this element of the method, with some researchers reporting a possible 2–3 per cent reduction in cost when using project partnering whereas others suggest savings of 60 per cent when practising strategic partnering.

This latter figure was reported by Westminster University [11], based upon the case study of the McDonalds construction programme for their standard, fast-food restaurants and reflects not only the use of strategic partnering but also the use of modular buildings and the repetitious nature of the projects and should therefore be viewed in this context.

The former figure results from research carried out by the New South Wales Department of Public Works and Services [6], which compared costs on ten individually partnered projects with ten non-partnered public projects. Other case studies of partnering in Australia, although not providing specific examples of cost savings, confirm that the use of the method results in a lower risk of cost overruns.

In the USA, a quantitative study [12] of 400 public projects, half of which were performed using project partnering, carried out by the Texas Department of Transportation found that whereas the partnered projects incurred more change orders (variations) than non- partnered projects the average cost was approximately one-half of that expended on non-partnered projects. The partnered projects had slightly less cost growth than non-partnered projects, and for the majority of the partnered projects no costs were incurred from claims or disputes. Because of the nature of the research, no attempt was made to identify any cost savings that may have resulted from these benefits.

The University of Reading's 1995 report [10] on partnering maintained that cost savings of 2–10 per cent were achieved on the project partnering schemes that were examined and that savings of up to 30 per cent could realistically be achieved when strategic partnering is used. The cost of implementing the method was put at around 1 per cent of the project cost.

Bennett and Jayes, in their 1998 report [5] on second-generation partnering, i.e. long-term/strategic partnering, maintained that where cost reduction was the focus of the exercise savings of up to 40 per cent were obtained. The cost of using partnering was not quantified, although the fact that costs will be incurred by all of the partners was acknowledged.

While the majority of the literature supports the contention that – when using both project and strategic partnering – cost savings can be achieved, the quantification of these savings has so far been based on too few projects to enable robust results to be obtained. As for the cost to the partners of implementing these methods, even fewer data are available, or are not sufficiently reliable, to enable specific, dependable, quantified conclusions to be drawn.

Speed

The 1995 Reading Construction Forum report [10] on partnering maintained that the benefits derived from partnering included reduced design times, quicker commencement on site and shorter construction periods. A 27 per cent reduction in construction times achieved on five projects carried out by the Arizona Department of Transport is given as an example.

In the Westminster University report [11], for two of the five cases studied dramatic reductions in construction and delivery times were achieved, substantial shortening of the construction period was achieved in one case study and in the remaining two case studies researchers were satisfied that although reductions were not made delays would have been worse if partnering had not been used.

An overview of the performance of partnering in the Australian construction industry [6] is more circumspect in its findings, which, in essence, simply confirmed that there was a lower risk of time overruns on partnered projects.

An analysis of partnered project performance carried out on 400 projects in the state of Texas in the USA was more positive, establishing that time growth was negative on all of the 200 partnered projects, which resulted in them being completed, on average, some 4.7 per cent earlier than originally planned.

Bennett and Jayes [5] maintain that when using second-generation/ strategic partnering, and provided that the firms involved carry out the correct management actions, savings in time of more than 50 per cent are achievable.

While there is little doubt that substantial savings in time have been made in specific circumstances, usually where strategic partnering has been used by organisations that were determined to make it work, there is again insufficient evidence available to support the suggestion of a very definitive range of percentage savings of time. It can, of course, be said with some certainty that the use of project partnering will reduce the possibility of time overruns and that strategic partnering may well produce significant reductions in both design and construction periods.

Quality

Three of the five clients studied in the Westminster University report [11] claimed to have achieved improvements in construction quality, although the most clearly observable increase resulted from the use of modular buildings which were constructed under factory conditions.

The Australian experience [6], although lacking in definitive data with regard to quality, confirmed in general terms that the use of project partnering resulted in an improvement in quality standards.

According to the Reading Construction Forum's report on partnering [10], much of the literature examined for the report suggested that the primary focus of partnering should be quality. Some of the case studies investigated showed a reduction in the number of defects identified, together with a related decline in the amount of remedial work needed.

Although the above report did not identify any dramatic improvement in the level of quality, the same organisation's subsequent publication [5] found that when using second-generation/strategic partnering there were no defects present at handover. When the next level of 'third-generation' partnering was reached, the quality of service given by all of the partners and the quality of the final product were improved even more.

Therefore, there appears to be little doubt that partnering in general will result in some improvement to the overall quality of both the service and product aspects of a project or construction programme, but there is very little quantitative evidence available to define this improvement more accurately. Any improvement is most likely to be achieved by the implementation of total quality management techniques as a joint exercise by all of the involved parties.

Other characteristics

Contractor's advantage

With few exceptions, the literature describing case studies of projects in the USA, Australia and the UK confirms that working relationships between all of the participants had improved as a result of the partnering process and that the number of unresolved disputes had been reduced, often by a substantial margin and very often by judicious resolution of the dispute when it arose. The level of conflict had been reduced, and communication and the level of teamwork among the various parties had been substantially improved when compared with non-partnered projects, as had safety standards.

One of the principles of partnering is that all of the members of the project team share in any benefits that accrue from using the partnering approach, e.g. improved productivity and profitability, innovation and, particularly, cost savings. However, anecdotal evidence from Australia suggests that such benefits are rarely achieved [6]. On the other hand, research based on case studies in the UK found that significant benefits can be enjoyed [5,10].

The fact that continuous improvement in the performance of the individual participants is one of the main aims of any partnering arrangement means that the efficiency of the partnering organisations improves not only in respect of the specific projects in which they are involved but also as a result of the transfer of such improvements in personal and corporate terms to the wider activities of the organisation.

7.4 Summary

Partnering has been used in the UK for a comparatively short length of time, and while the characteristics of the method have been well defined in the literature the quantitative effect that these have on the three primary project objectives of time, cost and quality/functionality have yet to be determined over a sufficiently long period to enable them to be considered robust enough to allow incontrovertible conclusions to be drawn.

The fact that research into the cost, time and quality aspects of the system in the USA, where the method has been used over a longer period on a greater number of projects, appears to produce more conservative results than in the UK is somewhat concerning. The American results are based upon a quantitative analysis of data and on the production of mathematically based project performance indicators for the various elements within the three primary objectives. This would appear to indicate their robustness and has led to the advantages gained from the use of partnering in these three areas to be more cautiously assessed than in some UK literature.

Advantages

1   The client's project costs are reduced, although these are difficult to quantify accurately as there is a very wide range of percentage savings

reported worldwide. There is little doubt, however, that strategic partnering will produce greater reductions over time than partnering on single projects.

2   Construction periods are reduced and design periods are sometimes shortened; again, the range of reported percentage reductions is so wide as to make it difficult to determine an accurate average or range.

3   Quality of the final product is improved as a result of partnering and safety standards are heightened. Mutually beneficial total quality management schemes can be jointly implemented.

4   Conflict among all of the members of the project team, and particularly between client and contractor and between contractor and subcontractor/ supplier, is reduced, as are the number of disputes and claims.

5   Communication between all members of the project team is improved, and the establishment of mutual objectives ensures that the client's needs and objectives are known and understood by all of those involved.

6   The achievement of more efficient working and greater productivity that has been recorded as a result of the principle of continuous improvement will benefit all participants in their general activities as well as the specific project(s) with which they are involved.

7   When involved in strategic partnering, the participants should benefit from the increased amount of innovative thinking and research and development that can be carried out as a result of sharing ideas and pooling resources.

8   Contractors, when involved in strategic partnering, are able to rely on a proportion of their annual workload being assured for at least the duration of the partnering programme, and the costs normally associated with tendering for the amount of work in the programme will not be incurred.

Disadvantages

1   Additional costs are incurred by all partners as a result of the need to hold workshops and to train staff, and also as a result of the expenditure on extra management time, etc.

2   There is always the possibility that the client will not be able, or may not wish, to honour the arrangement to provide as much work as was originally envisaged, leaving the other partners with voids in their order books which they may find difficult to fill.

3   The relationships built up during the partnered project(s) can lead to cosy associations which do not encourage new ideas or a dynamic approach to the management of the project. Team members, when returning to their parent organisations, may find that there has been a hiatus in their career progression and that they have difficulty in acclimatising to the different culture.

4   Confidentiality can be compromised as a result of the opening up of

information that is normally restricted to the organisation from which it originated.

5 Maintaining the commitment of staff, and particularly senior management, especially during the early stages of a project or programme when the benefits of partnering have not yet become evident, can be problematic and may even prove to be impossible.

6 Difficulties can arise as a result of conflicts that may occur between the partnering charter/framework, which usually are not intended to have any legal standing, and the terms and conditions of the formal contracts between the various parties involved in the project(s).

That partnering is currently popular with many major clients and certain sectors of the construction industry cannot be denied, but the lack of robust, long-term, quantitative data on the system's performance in the UK means that firm conclusions cannot yet be drawn as to its general efficacy.

Despite this, there seems to be little doubt that its use will, at least in the immediate future, continue to grow, as evidenced by the publication in September 2000 of the Association of Consultant Architects' Project Partnering Contract designed to establish new working relationships between all members of the project team through an integrated project process.

The advent of this contract, which is intended to eliminate the need for any further principal legal documentation, may well signal the transformation of project partnering in practice from an administrative framework into a fully fledged procurement system.

In addition, the fact that during the last years of the twentieth century central government through the Treasury and Defence Estates pioneered the use, initially on an experimental basis, of supply chain management in the form of a new system of prime contracting appears to indicate that such approaches will become part of best practice within the industry.

Prime contracting itself is not examined in this work as it is considered too early in its life to enable any worthwhile conclusions about its performance to be drawn, but it can be briefly described as a system in which the client has a single contractual relationship with the prime contractor, who is usually an established main contractor. The prime contractor is in control of a team of designers, specialist contractors and suppliers, all of whom have a long-term contract with the client for a programme of building projects. The project team works together with the client to produce the design and to implement the project within agreed non-variable cost, time and quality criteria; during the project's life the principles of partnering are applied.

References

1 British Property Federation (1983) *Manual of the BPF System for Building Design and Construction*, London: British Property Federation.

2   Royal Institution of Chartered Surveyors, Quantity Surveyors' Division (1984) *Report of a Working Party on the British Property Federation System for Building Design and Construction*, London: Royal Institution of Chartered Surveyors.

3   Construction Industry Institute (1991) *In Search of Partnering Excellence*, Adelaide, Australia: Construction Industry Institute.

4   Construction Industry Board (1998) *Fact sheet on Partnering*, London: Construction Industry Board.

5   Bennett, J. and Jayes, S. (1998) *The Seven Pillars of Partnering. A Guide to Second Generation Partnering*, London: Thomas Telford.

6   Uher, T.E. (2000) 'Partnering performance in Australia', *Journal of Construction Procurement* 5 (2), 163–176.

7   National Economic Development Council (1991) *Partnering: Contracting Without Conflict*, London: HMSO.

8   Davis, Langdon and Everest (2000) *Contracts in Use. A Survey of Building Contracts in Use During 1998*, London: Royal Institution of Chartered Surveyors.

9   Walter, M. (1998) 'The essential accessory', *Construction Manager* 4.

10   Reading Construction Forum (1995) *Trusting the Team. The Best Practice Guide to Partnering in Construction*, Reading: Centre for Strategic Studies in Construction.

11   Barlow, J., Cohen, M., Jashapara, A. and Simpson, Y. (1997) *Towards Positive Partnering. Revealing the Realities in the Construction Industry*, Bristol: The Policy Press.

12   Gransberg, D.D., Dillon, W.D., Reynolds, L. and Boyd, J. (1999) 'Quantitative analysis of partnered project performance', *ASCE Journal of Construction Engineering and Management* 125 (5), 161–166.

8   Common variants of the main procurement systems

8.1 Introduction

Over time, the well-established procurement systems, particularly the conventional method and design and build, have produced variants which are common to some, or all, of these systems and have evolved as a result of clients wishing to change, and thus improve upon, the basic methods in order to satisfy their own particular project needs. In many cases, this was achieved by varying the way in which tenders or proposals were obtained from the supply side of the industry. This chapter attempts to provide an overview of the various methods and their relationship with each of the procurement system categories.

8.2 Common variants

Two-stage selective tendering

When using this approach, tenders are invited on the basis of approximate, or limited, project documentation, and the successful first-stage bidder(s) (there can, especially where there is an element of contractors' design, be more than one) is/are asked during the second stage to collaborate with the client to produce a definitive design and agree a final tender figure.

Where more than one tenderer is involved in the second stage, the most appropriate bid is accepted and all of the involved organisations are usually reimbursed their second-stage costs. Two-stage tendering can be used with any of the systems within the separated or integrated categories, although the basic method may well have to be varied to suit a particular procurement path.

Once construction starts, the project proceeds in a similar manner to that adopted when using the basic chosen procurement system, with the issue of interim and final valuations and certificates being determined by the system's usual procedures.

Shorter preconstruction times have been experienced when using this variant, the average time overrun incurred is usually shorter than when using many of the other methods and there is less variability in performance in meeting time targets. Most authorities agree that a cost premium is usually paid for adopting this approach and that the need for price certainty should be a secondary consideration.

However, the fact that completion can generally be achieved earlier than when using the basic form of the system may well help to offset the additional cost that is likely to be incurred. It should also be recognised that it has been demonstrated that the use of this variant generates a very low value of variations, a phenomenon that probably reflects the co-operative nature of the method.

Some concern has been voiced about the possibility that, once initially selected, the sole tenderer can increase his/her level of pricing. However, where the selection process has been properly managed and documented, where a reliable basis of pricing has been established and where no significant changes in the client's brief or design concept have been requested, this problem should not occur. It has also been suggested that approximate documentation can be, and indeed has been, used as a palliative by design teams who have fallen behind schedule to overcome, or hide, their own inefficiency. While this may be a possibility, it is unlikely to be a problem on a well-managed project where all aspects of progress are strictly monitored by the client's project manager.

There is also a more general suspicion that approximate, or notional, bills of quantities and schedules of rates allow tenderers to weight their tender prices, i.e. in areas where they consider the quantities may eventually be substantially increased, although this sort of price-rigging should easily be detected by an experienced quantity surveyor during the detailed evaluation of the tenders.

When considering the use of this procurement method, it should be appreciated that, like most methods involving co-operation with a contractor, two-stage selective tendering calls for a greater input from the client, or his/her advisors, than a more conventional approach and that it is essential that all of the participants in the scheme are conversant with every aspect of the method in order to avoid any misunderstandings occurring during the currency of the project.

It is also necessary to ensure that mutually acceptable procedures are established by the client's representatives and the contractor in order to govern the methods by which the price for the project will be calculated and that, as a means of ensuring that value for money is being obtained, valid comparisons are made with similar projects that have been awarded on a competitive basis.

Despite these caveats, the majority of authorities agree that the use of this subvariant can be an appropriate solution where it is desirable to secure the early involvement of a contractor to provide some form of expertise, where the client wishes to commence work on site in advance of the detailed design of the project having been finalised, where it is possible to overlap design and construction and where the client wishes to minimise the overall project period.

Negotiated contracts

It is possible when using this variant to appoint a contractor by assessing the experience, management expertise and competitiveness of a small number of appropriate organisations. More commonly, an appointment of a single contractor is made on the basis of past performance and competitiveness on an identical, similar or geographically adjacent project, preferably carried out for the same client.

In the first method, detailed discussions are held with each of the chosen contractors, during which their experience and management expertise is assessed and the organisation offering the experience, skills and knowledge best suited to the project is selected. This enables the chosen organisation to participate in discussions with the client, or his/her consultants, to provide advice on buildability, value engineering and construction methods and eventually agree a price and/or a design for the project that will then form the basis of a contract, which will be entered into before construction work commences on site.

When – in the second method – negotiation is restricted to one contractor, the same process is followed, with the price for the project being established on the basis of the bills of quantities (or some similar financial document), or the rates contained within the bills, for a similar project.

There is little doubt that the use of negotiation can result in some saving of time during the preconstruction period, primarily as the result of the reduction of the period required for competitive tendering and when using any of the systems within the integrated category by adopting the contractor's design proposals.

Research carried out during the preparation of *Faster Building for Industry* [1] established that progress tends to be faster if a single contractor is chosen and a price negotiated with that contractor rather than obtaining competitive tenders, although the reasons for this phenomenon were not identified.

On the question of cost, most authorities are of the opinion that the best economical financial solution is rarely attained, and some believe that a premium is invariably paid when using this procedure.

It is always possible that negotiations may break down and the two parties reach deadlock, and it is thus necessary to ensure that the client has the option of being able to break off negotiations if the impediment to an equitable solution cannot be removed. In order to offset these financial disadvantages, it has been suggested that as the contractor can be made a party to the cost planning of the project, as a result of his/her involvement during the early stages of the contract, there may well be a hidden cost benefit to the client resulting from the input of the contractor's commercial expertise.

The conditions under which the use of the negotiated contract is appropriate echo, to a large extent, those already identified for the two-stage tendering approach, i.e. where the client wishes to achieve a modest saving in the overall project period, where any early start is required on site, where the contractor needs to be selected during the early stages of the project so that he/she can provide special expertise, where conventional competitive tendering cannot attract sufficient tenders or produce realistic prices and where the contractor is already on site and the cost of the site establishment, etc. may be saved or reduced.

As with the two-stage tendering system, the client should only deal with contractors who have previously demonstrated their reliability and managerial and technical competence and who are prepared to foster a relationship in which mutual trust and respect are paramount characteristics.

Continuity contracts

When using this variant, contractors bidding for a project on the basis of single-stage selective tendering are advised that the successful tenderer, subject to satisfactory performance, will be awarded a similar project to follow on from the completion of the first. The price for this subsequent project will be negotiated using, as a basis, the tendered rates included in the bill of quantities, or some other form of financial schedule, for the original project.

It is therefore a prerequisite for the use of this system that there are at least two similar projects available within a defined geographical area that can be carried out sequentially and that are capable of being able to accommodate flexibility in the timing of the commencement and completion of the second project. Opt-out clauses for both the contractor and the client are often included, even if all the criteria for success are met. Also, the criteria for measuring the success of the first project must be agreed by both parties to the contract, and procedures for negotiating the second contract must be established before the initial project is let.

With this variant, the guarantee of continuity does generally result in a more positive commitment from both consultants and contractors on the first project. The method has normally been used in conjunction with the conventional system of procurement, although in theory the process could be applied to any project irrespective of the procurement approach adopted.

When using this variant, it has been found that time overruns are shorter than average, cost overruns are more predictable than average, very competitive rates can be obtained at tender stage, the value of variations is likely to be low in comparison with other methods and few variations were needed on the second, or succeeding, projects. Some additional risk is taken by the client when using this variation of the parent system as there is a commitment to a second contract with no guarantee that the contractor will act as he/she did on the first project.

Serial contracts

In this method, a number of projects – often referred to as a programme – with similar characteristics, particularly in the case of the design of the development, are awarded to a single contractor following the receipt of competitive tenders based upon a master set of documents, the nature of which will be determined by the procurement method that is being used.

Although forming part of the same programme, each project is administered by means of a separate contract, with the contract sum for each being calculated by using the prices contained within the master tender and the amount of work appropriate to each project.

Serial contracts can consist of a number of projects, either with individual start and finish dates or with flexible timing to give continuity of work. Parallel working on different projects is quite common and has obvious advantages in terms of savings in cost and time.

The method was originally instigated by various central and local government bodies at a time when construction resources were in short supply in an effort to eliminate, or at least reduce, the inefficiency of allowing the knowledge and expertise of the project team that had been built up over the duration of a project to be dispersed as soon as the work was completed.

During the 1960s and early 1970s, this variant was successfully employed on projects where building systems such as CLASP and SCOLA (Second Consortium of Local Authorities) were used and on local authority housing schemes, school programmes and some central government building projects of a repetitive nature. The performance of this method was surveyed during preparation of the Wood Report [2] and, compared with the conventional procurement system, was found to be outstanding in terms of all three of the usual criteria of time, cost and functionality/quality.

The advantages of the method have been identified by Ashworth [3]:

1   When using the conventional procurement system, the contractor's long-term association with the client and his/her consultants and the experience gained of the types of work included within the programme permits close collaboration between designers and contractors and fosters a spirit of co-operation among all of the parties.
2   Improvements in the performance of all of the participants can result from the operation of the learning curve after completion of the first project; and continuous feedback from site to the contractor's senior management and the designers also leads to a more efficient use of resources and increases the buildability of the project.
3   Project teams can be moved to successive projects, thus avoiding the need – as with most other procurement systems – to dismantle and reconstruct these experienced groups.
4   Small client organisations with similar building requirements are able to amalgamate their building programmes and form consortia in order to let contracts on a serial basis.
5   Contractors are provided with continuity and thus security of workload.
6   Contractors' costs, and therefore tender prices, should be reduced as a result of higher rates of production achieved on site and the bulk-buying opportunities presented by the involvement in such building programmes.

Provided there is a firm programme of similar projects – all of which are located within a reasonable distance of each other – and the design process is well disciplined, serial contracts can be used on all types and sizes of appropriate projects. The use of this variant is enhanced if the building programmes can be linked with large-scale production of factory-made components or bulk purchasing of materials, equipment and products.

Where clients have an ongoing long-term programme of work, even if the projects are not all similar in nature, this procurement method has been found to be extremely effective in reducing costs and programme times without detriment to quality. Long-term partnering arrangements enable this method to be used to the benefit of all of the participating parties.

This method of obtaining buildings supports the current trend for moving away from approaching project procurement on a project-by-project basis and can result in an increase in innovation and productivity as well as improvements in communication and a reduction in the confrontational attitudes that still pervade the industry.

Cost-plus contracts

Under this arrangement, a contractor is appointed, usually on the basis of competition on the fee element of the project only, to carry out the work defined by the client or his/her consultants, with reimbursement being made by the payment of the actual (prime) cost of the works and a fee to cover the contractor's overheads and profit.

The contractor's fee can be calculated in a number of ways and can be:

1   a fixed fee, in the form of a lump sum based upon the estimated cost, that is varied only if the nature of the project changes dramatically;
2   a percentage fee calculated on the final cost of the project;
3   a percentage fee, related not to cost but to the estimated value of the project at the outset, updated by any variations that occur during the currency of the work.

Target-cost contracts

This method differs from the basic cost-reimbursable variant in two main aspects, both of which affect payment.

First, a contractual agreement is reached, either in competition or negotiation, on a target cost for the work and a fee to cover the contractor's overheads, management costs and profit. Second, a procedure is agreed for sharing any savings or additions if the actual cost is lower, or higher, than the target costs. This mechanism provides a financial incentive to the contractor which is absent from the basic method.

Targets may be applied individually to all of the principle elements of a project, i.e. cost, time and functionality/quality, with the degree to which the target is met by the contractor being reflected in the eventual reimbursement received by him/her. It is also possible to set targets for two or all of the three elements and link them to form a combined target.

General

The absence of a tender sum or even an estimate of the cost of the final account and the fact that the appointed contractor has little incentive to control his/her expenditure means that this method is only used by clients under certain circumstances and not for normal run-of-the-mill projects. However, when using this method, there is little likelihood of contractual claims for additional reimbursement being submitted by the contractor.

The financial and practical risks associated with the project are shared more equitably than when using most other procurement methods, the extent and speed of the work can be adjusted to the availability of project funding and the client's total financial commitment can often be favourably compared with that for conventionally managed projects that have been subjected to extensive claims.

Conversely, the cost of rectifying the contractor's mistakes, or inefficiencies, together with matters which under other circumstances would become claims, are often hidden and eventually charged to the client notwithstanding that the responsibility often lies with others. This characteristic and the fact that there is no contractual commitment by the contractor to the final cost to be paid by the client or direct financial incentive for him/her to make the most efficient use of his/her resources means that the method cannot easily satisfy the concepts of public accountability.

Experience has shown that negotiations with the contractor during the early stages of the project can be protracted – the process of establishing the actual cost of the work can be cumbersome, onerous and difficult to monitor, the lack of incentive for the contractor and his/her workforce to perform well necessitates much closer client control than is normally required and the fixing of targets, should they be used, requires very fine judgement.

The advantages associated with this method include the ability to commence construction with the minimum of design having been completed, with the consequent flexibility to develop the design, the programme and the scope of the work after construction has started together with the availability of proven forms of contract for the variant. Early contractor involvement is of course inherent in the method, enabling construction expertise to be introduced during the design stage and allowing the contractor to be better able to plan his/her on-site construction activities.

It will be apparent that the use of cost-reimbursable contracts should be strictly controlled and only implemented in emergencies or where the use of all other procurement systems has been examined and discounted and a combination of the following circumstances prevail:

1   where the client must resume or complete a project disrupted by strikes, bankruptcy or other influences and in emergencies;
2   where major unquantifiable risks are present;
3   where market conditions are so buoyant that normal tendering procedures are not appropriate and inflation is rampant;
4   where the project is of exceptional managerial and technical complexity;
5   where the scope of the works cannot be readily defined at tender stage.

Term contracting

This method is used when bids need to be obtained for work that, although of a minor, repetitious nature such as building repairs and maintenance, is continuous for the life of the buildings and incurs substantial annual expenditure. Bids are sought from suitable contractors and are usually based on specifications and comprehensive schedules of rates, which cover all of the major items of work likely to be carried out during the length of the contract, which can vary from 1 to 3 or more years.

8.3 Summary

All of the variants enable some form of early collaboration to be achieved between the client and members of the project team. In the majority of cases, this takes the form of the contractor contributing to the design of the project and/or giving advice on buildability, costing, material ordering, programming, etc.

The majority of the variants will enable modest savings in time to be achieved at the expense of the final cost, which will generally be higher than when using the main procurement system itself. Many of the variants can and should only be used in special circumstances and need to be carefully and firmly controlled by the client.

The main advantages and disadvantages of the various variants are now summarised:

Two-stage selective tendering

Advantage

Savings in time can be achieved using this method, and where overruns are experienced they are usually shorter than in any other conventional method.

Disadvantage

Work is commenced before a final tender sum is agreed, therefore early price certainty needs to be a secondary consideration and the client can be vulnerable to any change in the level of the contractor's pricing from that contained within the first-stage tender.

Negotiated contracts

Advantages

1 Modest savings in time can be achieved using this method.
2 The system is useful where other procurement methods cannot attract sufficient tenders or realistic prices, where a special expertise is required or where project costs can be reduced as a result of the contractor already being established on site.

Disadvantage

A cost premium is invariably paid by the client when using this method, and the project cost is thus nearly always higher than when using other procurement systems.

Continuity contracts

Advantages

Very competitive rates and tenders can be obtained when using this method, the value and frequency of variations are lower than when using other systems and time overruns will usually be shorter than those experienced on projects managed by some other procurement methods.

Disadvantages

1 This system can only achieve its maximum potential if there are at least two similar projects available within a defined geographical area which can be carried out sequentially and are capable of accommodating some flexibility in the timing of the commencement and completion of the second project.
2 The client is committed to a second contract with no guarantee that the contractor will act, or perform, as he/she did on the first project.

Serial contracts

Advantages

1 This approach avoids the need to dismantle experienced project teams after the completion of one project and allows their accumulated knowledge and expertise to be utilised on the other projects contained

within the serial programme with the result that the method has proved
to be outstandingly successful in terms of the usual performance criteria
of cost, time and functionality.

2   Tender prices are able to be reduced as contractors are given continuity
of work and the ability to 'bulk buy' materials, particularly if the client's
building programme can be linked to the large-scale factory production
of components. This benefit can be even greater if different clients take
advantage of the opportunity that this system offers to amalgamate their
individual building programmes.

Disadvantage

This system can only be used if the client has a substantial and ongoing
building programme in which the individual projects are sufficiently similar
in design to enable master documentation and common tender/proposal
documentation to be produced.

Cost-reimbursable contracts

Advantage

When there is inadequate definition of the work at the time of tender, when
high inflation is prevalent, when the project is extremely complex, when
there is a major or unquantifiable risk or when an emergency occurs, the
use of this system can be advantageous to the client, provided that a number
of these characteristics are in combination within the one project.

Disadvantages

1   The absence of a tender sum and an estimated final cost generally
precludes the use of this system on projects that are subject to rigid
accountability requirements.

2   There is no contractual commitment by the contractor to the final cost
and no financial incentive to use his/her resources efficiently. Although
incentives can be incorporated to mitigate this difficulty, the fixing of
targets requires very fine judgement.

The next chapter is devoted to the theory and practice of choosing the
most appropriate procurement system, or variant, to match the client's
organisational culture and needs and the project objectives. The
discriminating use of the information about the various procurement systems
contained within the previous chapters will be critical to the success of this
activity.

References

1 Building Economic Development Committee (1983) *Faster Building for Industry,* London: National Economic Development Office.
2 Building Economic Development Committee (1975) *The Public Client and the Construction Industries (The Wood Report)*, London: National Economic Development Office.
3 Ashworth, A. (1986) *Contractual Procedures in the Construction industry,* London: Longman.

9 The selection of building procurement systems

9.1 Introduction

In the previous chapters, the various procurement systems were identified and described, and it is from this wide range of means of procuring the design, construction and other aspects of the project that the client has to select the most appropriate method to ensure that his/her objectives are met.

In order to examine and understand how such selections should be and are made, it is suggested that it is necessary, first, to recognise the principles of decision-making and choice in general. The vast body of literature and research that has been generated over the past half-century on this subject makes the task of encapsulating these principles within a few pages, while at the same time avoiding an oversimplistic approach, a difficult one, and once again the reader is advised to read the specialist literature detailed in the references.

Before carrying out such an examination, it is necessary to determine an acceptable definition of the term 'decision'. *The Concise Oxford Dictionary* [1] defines 'decision' as a 'settlement (of question etc), conclusion, formal judgement; making up one's mind, resolve; resoluteness, decide character'. It is suggested that a simpler and more straightforward definition should be: 'The process of choosing one action from a number of alternatives'.

However, most authorities agree that the act of reaching a decision, no matter how the act itself is defined, is only the final stage in the dynamic process of decision-making. When definitions of the whole process are examined, a common theme can be extracted of a dynamic process stimulated by the identification of a need and culminating in the act of making a choice between alternative means of satisfying such a need.

The examination of decision-making that now follows is therefore carried out by looking at the process in its entirety rather than just the final act of reaching a decision.

9.2 Theoretical decision-making and choice

The decision-making process

It has been argued, and accepted by most authorities on the subject, that although decision-making can take many different forms and can be carried out by individuals, groups and organisations the process and stages within it will remain constant and have certain common features. Harrison [2] maintains that the decision-making process is straightforward, made up of seven steps and largely sequential, with its normal progression being:

1   to set organisational objectives;
2   to search for alternatives;
3   to compare and evaluate alternatives;
4   to choose among alternatives;
5   to implement the decision;
6   to follow up and control by taking corrective action if necessary;
7   to revise and update objectives.

The process in this model is circular and dynamic, although in the case of simple uncomplicated decision-making stages 1–5 could be used in isolation without detriment.

The common elements within and characteristics of theoretical decision-making are therefore that:

- Decision-making is a circular/continuous activity, although in the case of the selection of a procurement system for a single project the process normally only lasts for the duration of the project.
  Decisions made regarding design and many other aspects of the project will, however, have an effect on the management of the facility for the whole of its life and thus be part of a continuous process.
- Most, if not all, organisations have a continuing requirement to establish objectives in order to meet their ever-changing needs and, in the process of attempting to meet these objectives effectively, a variety of decisions must be made.

  The decision that concerns us here is the selection of the most appropriate procurement system to satisfy the client's needs related to a specific project.

- To assist in making such decisions, the use of a framework, which enables the organisation to move through the decision-making process in an orderly and efficient manner, must be helpful.
- The framework will normally consist of a series of simple steps, whereby the problem is identified and diagnosed; objectives are then set and

constraints considered; alternative solutions/courses of action are identified; a choice is made and the consequent decision implemented.

A number of such frameworks, which have been designed to assist in the selection of the most appropriate procurement system, are described and illustrated later in this chapter.

- Once execution has taken place, or is continuing, the activity is reviewed in order to establish whether the original problem identification, investigation and choice of solution were correct or whether further remedial action is necessary.

In the case of procurement system selection, once the choice has been made and implemented it is unlikely that any remedial action will be practically possible as any attempt to change the method of procuring the design and construction will be too costly and will result in delays to the project. This situation reinforces the need to delay the selection until the latest possible time and to ensure that every effort has been made in the light of all the available information to ensure that the correct choice is made.

Whilst the decision process framework would probably be seen by most authorities as an acceptable – if somewhat simplistic – view of the theoretical decision-making process, it is at this point that the theorists have, in the past, parted company. Their differences as to how decisions are actually made within this setting are epitomised by what has become, over the past half century, the opposed schools of rational, behavioural and political decision-making theory.

The rational classic approach stems from an impressive body of formal theory that has been erected since the end of the Second World War by mathematicians, statisticians and economists and is encapsulated in subjective expected utility theory. This work has been described as one of the most impressive achievements of the first half of the twentieth century [3].

The theory revolves around the concept of rationality and assumes that the decision-maker knows exactly what he/she wants; has an identifiable set of alternatives to choose from and is aware of their outcomes; is able to choose the alternative that gives the best value and is able to ignore any constraints within his/her environment.

A brief examination of these criteria soon reveals that, while the rational approach has validity where the decision-maker is in possession of perfect information and is able to meet all of the identified requirements, very few individuals or organisations operating in the real world are ever in such a state of certainty when making decisions. Usually, decision-makers are only in possession of part of the necessary information needed to examine all the alternative solutions and determine their consequences. Often, organisations have difficulty in accurately identifying the cause of the problem that is compelling them to enter into the decision-making process and are unable to produce even a simple list of alternative means of solving the dilemma.

Certainly, in the case of clients of the construction industry, their 'problem' is usually relatively easily identified – they require some form of facility that will enable them to provide a service or a product – but in most cases they are not in possession of all of the necessary information to allow them to make a rational decision as to how to achieve their objectives. In an attempt to deal with these practical difficulties, March and Simon [4] pioneered the alternative theory of bounded rationality, or satisfying, which asserts that decision-makers are hardly ever in a position to be fully rational, i.e. they operate under conditions of uncertainty and, in reality, reduce the complexity of decision-making by constraining the process of developing alternatives and processing information.

However, this approach assumes a perfectly unitary decision-maker, with information, knowledge and judgement shared among the responsible members of the organisation and with everyone agreed about the goals, values and perceptions; a situation which rarely, if ever, exists in any corporate establishment.

If, then, the rational/utility, or even the satisfying, theory of decision-making is too unrealistic in its assumptions, what alternatives exist among the behavioural school that might more accurately serve as a model?

The behavioural decision theory, first expounded by Cyert and March [5], was based upon an expansion of March and Simon's description of bounded rationality, but included their own ideas of the mechanics of the decision-making process that they believed were used in large organisations operating under uncertainty in imperfect markets.

This theory suggests that alternative solutions are not compared with each other, but are examined sequentially and accepted, or rejected, on the basis of target aspirations or their consequences. Where, during the decision-making process, there are any internal conflicts of interest, these are controlled by buffering, or isolating, inconsistent demands from the interested parties for lengthy periods. Rule-following, more than the calculation of the consequences of implementing alternative actions or solutions, determines most decisions, and decision-makers adopt strategies which avoid the ambiguity that exists within much of the decision-making process.

Although incorporating most of the substance of decision-making, the behavioural theory does not address the possibility that underlying any collective decision is a process of bargaining in which choices are determined by the resources committed by each member of the organisation to the achievement of some satisfactory solution.

The political bargaining model, formulated by Allison [6], supports this view and suggests that a process of bargaining between participants enables a final decision to be made which has the general support, and accommodates the interests, of all the participants.

While both the rational and behavioural approaches still have their immutable advocates, most authorities now agree that the decision-making process is an amalgam of all three models, although agreement on the relevant importance of each is unlikely to be reached [7].

Organisations operate in an environment in which external sources such as the country's political institutions, culture, traditions, laws and social mores exert pressures of varying kinds upon the decision-maker. Where choices and decisions are made without cognisance of any element of the environment, they are unlikely to be capable of being implemented [2]. The organisational decision-maker is also affected by the economic systems within which the organisation operates, i.e. the economic environment, which includes customers, competitors, the supply chain, associated industrial organisations as well as local and central government.

The effect of the economic environment is often overlooked when choosing a procurement system and tendering strategy. When contractors, subcontractors and suppliers order books are full, they will only be tempted to submit bids when the method of procurement is attractive, risks are minimised and it is likely that the successful tenderer will achieve an acceptable level of profitability on completion of the project.

The actions of government are particularly relevant to the decision-making process. In order to ensure that that the country can be fairly and effectively governed to the general benefit of its citizens, that the economic system is properly regulated and that relationships between parties are equitable, legislation needs to be enacted. Unfortunately, the laws and acts stemming from this need place a set of restraints upon the decision-maker within which choices must still be made in order to meet the organisation's objectives.

The decision-making process is also influenced by the social fabric of our world; the forces of social change and the plethora of social problems exert continuing and increasing pressure on corporate decision-makers. Irrespective of other influences that may affect the environment in which any decision has to be made, these problems increase the level of inherent uncertainty and risk already present. The economic system is not only concerned with what, and how many, goods and services will be produced but also with how, when and where these will be exchanged [8].

It can thus be seen that, in theory at least, the decision-maker – in addition to being in a constant state of uncertainty – must deal with the idiosyncrasies of his/her own organisation's structure and politics and at the same time take into account the economic, social and legal consequences of his/her eventual decision.

9.3 Decision-making in the real world

It is now intended to look at how decisions are actually made in industrial and commercial organisations and, where appropriate, to compare theory and practice.

The vast amount of research and literature describing normative decision-making is legendary, but there are relatively few hard data about how organisations make decisions [9]. There is, of course, a body of literature describing work on practical decision-making that has been carried out since the 1950s, but an examination of this reveals that there has been little in the way of systematic comparisons of large numbers of decisions until the Bradford studies [10], which began in the early 1970s and continued for over a decade thereafter.

As we have seen in the classic ideal, decision-making is seen as a series of logical steps with clearly perceived goals and with the ability to conduct comprehensive searches and obtain complete information; in other words, an orderly well-disciplined process.

When dealing with the actual making of what are often called *programmed* or *structured* decisions, this view has some validity, as such decisions are routine in nature [11,12]. Established organisational procedures, often based upon past experience, exist to deal with them and the solution is often obtained by computing or applying formulae.

In construction, such decisions are regularly made, e.g. when carrying out development appraisals and planning the implementation of simple repetitious projects where there is ample information available on the technical and managerial aspects and where the client organisation, and its advisors, are familiar with the type of project and are able to determine the project's parameters in the knowledge that few, if any, changes will be made through the duration of the work.

However, when describing the making of strategic decisions within organisations, most authorities agree that these are non-programmable *unstructured* processes. Such decisions, when they are made within organisations, are for the most part incapable of being programmed owing to their complexity, ambiguity, susceptibility to external influences and lengthy gestation period. Strategic decisions are associated with the long-term management of the organisation and concern the goals to be pursued and the strategies needed to achieve them.

A strategic decision has a number of characteristics that differentiate it from other lesser decisions. It will be comparatively novel, being more rare and non-routine than most; there will be few precedents for it, but it will itself set precedents for future decisions; it will commit substantial resources; it is often organisation-wide in its consequences and triggers other lesser decisions, and is thus atypical, important and all pervading [2].

A long-term programme of new-build construction and the implementation of most major projects, and indeed the selection of the most appropriate procurement method to use for the implementation of such developments, exhibit many of these characteristics and can therefore be treated as strategic decisions.

Strategic decision-making is a disorderly process and in reality does not conform to the ideal phases of decision theory, with phases being skipped, taken out of order and generally appearing to be ill-defined. Notwithstanding all of this, decision-making is a fundamental part of organisational life and needs to be managed.

The majority of strategic decisions appear to arise from deliberate managerial strategies which reflect the organisation's aims and objectives, although difficulty is often experienced in determining what the problem/opportunity actually is and/or setting objectives to deal with it. Most of the action in the decision-making process takes place when alternative solutions and/or courses of action are being identified and developed. Conflict can occur during this lengthy phase as information and opinions are obtained and as alternatives identified and considered [13].

The data obtained from the Bradford studies established that the organisation within which the decision is made influences and shapes the decision-making process by determining what are acceptable topics and by laying down the rules of the game.

Uncertainty is present at the evaluation stage in great abundance as a result of the lack of knowledge of the consequences of the various alternatives being considered. Under these circumstances, decision-makers, while using as much data, experience and beliefs as they can muster, often fill some of the gaps with hope and expectations.

The choice process, which should consist of the evaluation of alternatives, screening by the use of secondary constraints and determining which alternative enables primary goals to be best met, is distorted by a surfeit of information, bias, change, politics and internal and external interest. More difficulties are experienced during the authorisation of the choice than in the evaluation phase as the decision often needs to be considered within a limited time-scale and in the light of other contemporary decisions, with the authorisers often lacking sufficient detailed knowledge of the proposal and sometimes being subject to external pressures.

The decision-making process in general is not the classic model of problem-solving following a logical sequence beginning with the problem being defined and continuing sedately and smoothly through scanning information, assessing alternatives and choosing a solution. In practice, then, the process usually begins with very little understanding of the decision situation, a vague notion of possible solutions and very little idea of how to evaluate them and choose the most suitable solution. It is recursive, discontinuous and ever-changing; it also takes place over a considerable period of time and derives from a combination of its organisational setting and the problems and interests associated with the matter of choice.

9.4 The theoretical selection of building procurement systems

The procurement system selection process is, of course, a decision-making process, and Cole's [14] decision model is relevant to, and descriptive of, this process (see Figure 9.1).

Herman's [15] 'organisational iceberg' concept is also relevant to the procurement system selection process, reinforcing as it does the powerful influence of the covert criteria of perceptions, knowledge base, group confidence, structure of industry, attitudes, feelings, values/norms and personal interactions.

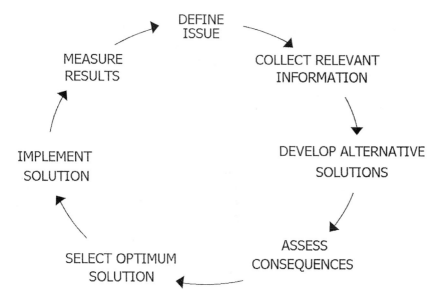

Figure 9.1 Cole's decision model

Unfortunately, very little research appears to have been carried out which specifically relates to the theoretical making of decisions in the construction industry or to choosing building procurement systems, with the majority of the literature relating to theoretical guidance and aids in the form of charts and computer programmes. This and other relevant literature are now examined on the assumption that the fundamental decision as to whether there is a need to build in the first place has always been properly established [16].

The construction industry is a fragmented diverse industry responsible for the design and construction of unique and often extremely complex, expensive and lengthy projects carried out in a hostile environment by means of what is usually a temporary management organisation consisting of groups of highly qualified individualistic professionals who are frequently unknown to each other.

The environment in which projects are implemented is thus far removed from the relative stability, permanence and sense of company loyalty which usually exists in industrial and commercial client organisations. It is generally accepted that, as a prerequisite to the implementation of any project, the client – particularly the inexperienced client – should have an understanding of the workings of the construction industry, its procedures and the characteristics of temporary management organisations as well as having some overall knowledge of all of the available procurement systems.

The Construction Round Table's publication *Thinking About Building* [17] was designed for this purpose and suggests the use of 'seven steps to success' when planning a building project:

1   appoint an in-house project executive;
2   appoint, if required, a principal adviser;
3   carefully define the project requirements;
4   realistically determine the project timing;
5   select an appropriate procurement path;
6   consider the organisations that will be employed;
7   professionally appraise the proposed site/building before making a commitment.

Assuming that the first two steps have been taken and that the client has an understanding of the construction industry, the third and fourth steps – defining the project requirements and determining the project timing – lead us back to the fundamental need to establish a clear and comprehensive brief. This should contain not only the aesthetic, technical and performance criteria for the project but also, of equal importance, the primary and secondary objectives in terms of functionality, quality, time and cost [18] and the political, cultural and economic environment within which the project will be implemented.

However, the definition of the project requirements needs to be broken down into two stages: first, a review of the client's needs, which requires that the construction industry must be capable of understanding organisations as well as managing design and construction, and, second, a definition of the means by which these needs will be achieved.

HM Treasury's Central Unit on Purchasing, in its guidance on *Contract Strategy Selection for Major Projects* [19], provides specific help in defining the means, in this case a six-part path, by which the selection of an appropriate procurement system can be achieved:

* *Review of contract strategy* – where an examination of the available procurement systems, project objectives and risks and responsibilities is carried out.
* *Analysis* – where a preliminary contract strategy is developed with a view to providing an initial indication of the constraints peculiar to the project and to consider how project objectives can best be achieved.
   Subsequently, a detailed strategy should be formulated, taking into account:
   – factors outside the control of the project team;
   – sponsors' resources;
   – project characteristics;
   – ability to make changes;
   – risk management;
   – costs issues;
   – timing; and
   – quality and performance.

- *Options* – where alternative solutions are developed and a recommendation for the selection of the most appropriate procurement system is made.
- *Selection of best strategy* – where the project sponsor is provided with the rationale and arguments for selecting the recommended system and, after discussion and perhaps some changes to or development of the proposal, finally approves the method of procurement which best satisfies the objectives of the organisation and the project.
- *Implementation* – where the project manager establishes the necessary resources, organisation, systems and controls required to implement the project.

Figure 9.2 shows the flow chart which illustrates the procedure in graphic form.

Mohsini and Davidson [20], in a paper describing the problems associated with the selection of the most appropriate procurement system, also illustrate the theoretical procurement route selection process by means of the flow chart shown in Figure 9.3.

The researchers maintain that, once a 'building' solution has been identified, the activities of the owner fall into the three blocks, A, B and C, as shown on the chart.

- *Block A* – activities in this block include an analysis of any past experience with the building process and recognition of specific objectives and constraints and enable the owner to decide whether to rent or purchase.
- *Block B* – activities allow the scope and requirements of the project to be identified and the project to be located so as to determine whether the facility can be acquired off the shelf or needs to be purpose designed.
- *Block C* – activities lead to the owner commissioning the various services needed to implement the purpose-designed project.

Within the activities described in block B, a decision would also need to be made as to the most appropriate procurement system for the project as the appointment of a design team and/or contractor should not normally be made before the best means of designing and constructing the development is selected.

Birrell [21] has suggested an 'improved research based' method of selecting the most appropriate building procurement system, with the choice being made by considering the client's objectives, the nature of the client, the state of the current local construction market-place as well as the typology of the project and matching of these to the most appropriate procurement system. Selection is achieved by means of an undefined system of 'votes' for the ability of the various procurement systems to satisfy the stated needs.

HM Treasury's guidance note on the *Selection of Contract Strategies for*

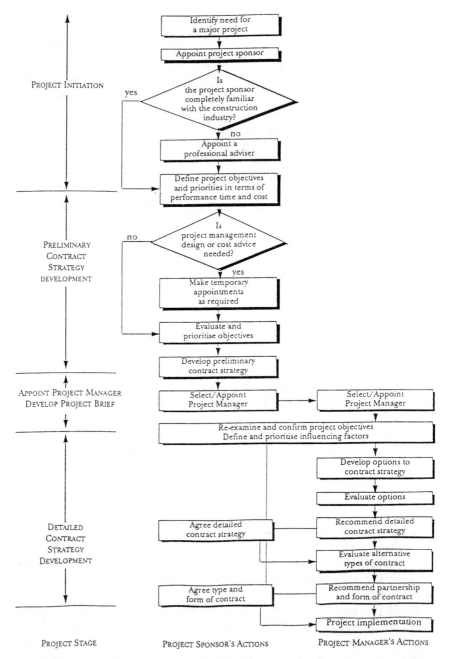

Figure 9.2 Contract strategy selection, after HM Treasury Central Unit on Purchasing [19]

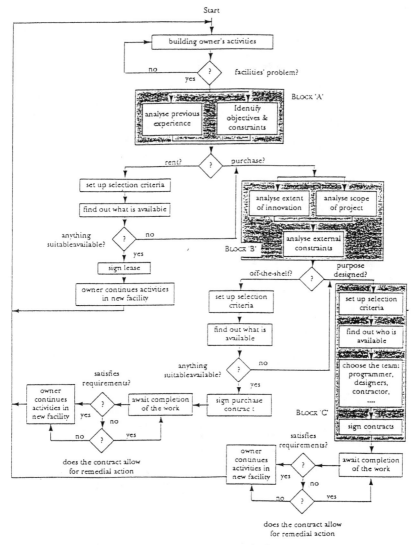

Figure 9.3 The building owner's strategic procurement decision process, after Mohsini and Davidson [20]

Major Projects [19] summarises the characteristics of the most common procurement systems (Table 9.1) and is intended to be an initial guide to the appropriateness of the various methods. In the same vein, *Thinking About Building* [17] provides a similar but more detailed chart (Table 9.2) that is also intended as a 'primer for discussion'.

The preceding methods of preliminary selection should be recognised for what they are: a crude means of reducing the possible alternatives to a

Table 9.1 Advantages and disadvantages of contract strategies, after HM Treasury [19]

Project objectives		Appropriateness of contract strategy in meeting project objectives				
Parameter	Objectives	Traditional	Construction management	Management contracting	Design and manage	Design and build
Timing	Early completion	□	■	■	■	■
Cost	Price certainty before construction start	■	□	□	□	■
Quality	Prestige level in design and construction	■	■	■	□	□
Variations	Avoid prohibitive costs of change	■	■	■	□	□
Complexity	Technically advanced or highly complex building	□	■	■	■	■
Responsibility	Single contractual link for project execution	□	□	□	■	■
Professional responsibility	Need for design team to report to sponsor	■	■	■	□	□
Risk avoidance	Desire to transfer complete risk	□	□	□	□	■
Damage recovery	Ability to recover costs direct from the contractor	■	□	■	■	■
Buildability	Contractor input to economic construction to benefit the department	□	■	■	■	□

Notes
■ appropriate; □ not appropriate.
This table is for guidance only. Generally, the appropriateness of the contract is not as clear-cut as indicated. The project manager must advise the project sponsor on this.

manageable number and the initial phase of the disciplined and systematic procurement system selection process.

The selection process has become increasingly complex, mainly as a result of the continuing proliferation of different methods of procuring building projects, the projects' ever-increasing technical complexity and the client's need for speedy commencement and completion, which has led to a demand for more sophisticated and systematic methods of selection to be devised.

The major difficulties in devising such methods have been identified by Skitmore and Marsden [22] as:

- No single person, or knowledge 'tsar', has been found who is fully conversant with all the main procurement arrangements.
- No consensus has been found between experts which easily systemises the procurement selection.
- Examinations of the factors affecting procurement selection have shown that no mutually exclusive sets of criteria uniquely and completely determine the appropriate procurement arrangements for a specific project.

Despite these strictures, methods have been devised by various researchers which partially overcome these difficulties or, in the case of the simpler approaches, ignore them entirely.

Franks [23] rates each of the systems that he describes in terms of their ability to satisfy the seven basic performance requirements, or expectations, that he has identified as being common to the majority of clients. Table 9.3 replicates his method of rating four systems. The ratings relate to a scale of 1–5, where 1 is the minimum and 5 is the maximum in terms of the individual system's ability to satisfy the listed requirements, and they are, of course, subjective.

While the inclusion of 'project management' among the list of procurement systems is considered ill-advised, the principle of the use of a method of rating the characteristics of the various methods against a list of client's needs is well established and has been used by HM Treasury's Central Unit on Purchasing [19] in their *Guidance Document No. 36*. This document suggests that the method often used for evaluating contract strategies (procurement systems) is by scoring how each system is able to meet the client's needs and project objectives, with these scores then being individually weighted according to their relative importance. The total of the weighted scores reflects the suitability of the various systems for the project under consideration.

The example shown in Table 9.4 illustrates the point that this process of evaluation can only provide guidance in selecting the most appropriate system; each option will have some disadvantage, or element of risk, associated with it but some will be better suited than others.

While, in the example, the traditional and design and build options appear to be the most suitable, the former is unable to satisfy the need for early

Table 9.2 Identifying your priorities, after Construction Round Table [17]

			Lump sum contracting		Design and build			Fee construction		Design and manage	
			Sequential	Accelerated	Direct	Competitive	Develop and construct	Management contracting	Construction management	Contractor project manager	Consultant project manager
A Timing	How important is early completion to the success of your project?										
	Crucial	1		•	•		•	•	•	•	•
	Important	2	•	•	•	•	•	•	•	•	
	Not as important as other factors	3									
B Controllable variation	Do you foresee the need to alter the project in any way once it has begun on site, for example to update machinery layouts?										
	Yes	4		•		•	•	•	•	•	
	Definitely not	5	•		•		•				
C Complexity	Does your building (as distinct from what goes in it) need to be technically advanced or highly serviced?										
	Yes	6		•	•	•	•	•	•	•	
	Moderately so	7	•	•	•	•	•	•	•	•	
	No just simple	8	•	•			•	•	•	•	•

			No.	1	2	3	4	5	6	7	8	9
D	Quality level	What level of quality do you seek in the design and workmanship?										
		Basic competence	9	•	•	•	•	•	•	•	•	•
		Good but not special	10			•	•		•	•	•	•
		Prestige	11		•		•					
E	Price certainty	Do you need to have a firm price for the project construction before you can commit it to proceed?										
		Yes	12		•	•	•	•	•	•	•	•
		A target plus or minus will do	13		•		•					
F.	Competition	Do you need to choose your construction team by price competition?										
		Certainly for all works contractors	14	•	•		•	•	•			•
		Works and construction management teams	15		•		•					•
		No, other factors more important	16			•	•			•	•	
G	Management	Can you manage separate consultancies and contractors, or do you want just one firm to be responsible after the briefing stage?										
		Can manage separate firms	17	•	•			•	•	•	•	•
		Must have only one firm for everything	18									
H	Accountability	Do you want direct professional accountability to you from the designers and cost consultants?										
		Not important	19	•		•	•		•	•	•	•
		Yes	20			•	•					
I	Risk avoidance	Do you want to pay someone to take the risk of cost and time slippage from you?										
		No, prefer to retain control and therefore risk	21		•						•	•
		Prepared to share agreed risks	22					•				
		Yes	23		•			•	•	•		
	Totals											

Table 9.3 Rating of procurement systems, after Franks [23]

Management system Client's performance requirements/expectations	Traditional	Management contracting/construction management	Package deal/design-and-build	Project manager/client's representative
(a) technical complexity; the project has a high level of structural mechanical services or other complexity	4	5	4	5
(b) high aesthetic or prestige requirements	5	3	3	4
(c) economy; a commercial or industrial project or project where minimum cost is required	3	4	4	4
(d) time is of essence; early completion of the project is required	2	4	5	4
(e) exceptional size and/or administrative complexity; involving varying client's/user requirements, political sensitivity, etc.	2	4	4	5
(f) price certainty; is required at an early stage in the project's design development	4	2	4	4
(g) facility for change/variation control by client, users or others during the progress of the works	5	5	1	4

completion and the latter is unable to maintain a direct link between the client and the design team. These weaknesses may well be overcome by using a variant of either of the two basic systems that exhibits the appropriate characteristics.

As in the majority of cases, when using this method of assessment, there is unlikely to be a clear-cut decision and managerial judgement will need to be exercised in order to determine the final choice. In this context, other factors – particularly the ever-changing political, economic and social environments – will also affect the decision-making process.

In 1988, Skitmore and Marsden [22] reported on their attempt to formulate a universal procurement selection technique through the use of two classic theoretical methods of decision-making and choice. The first of these was a multiattribute technique based on the original edition of the 'identifying your priorities' chart illustrated in Table 9.2, but modified to overcome what were seen by the researchers as two major deficiencies. These deficiencies were, first, that the answers to the various criteria listed in the original chart were restricted to a maximum of three alternatives. This was

Table 9.4 An example of contract strategy (procurement system) selection

Standard criteria	Project requirements	Requirement's relative weighting	Project manager's contract scoring			
			Traditional	Design and build	Management contracting	Construction management
Timing	The project must be completed within 30 months. The decision to proceed will be made within 3 months	4	4	10	8	8
Variations	The brief is well defined. It is unlikely that there will be major changes after construction commences	1	8	4	8	9
Project nature	The building design is similar to those constructed for other departments in recent years but somewhat larger	2	8	5	8	8
Quality	The specification for the building is of a relatively high, but not prestigious, standard.	2	8	5	8	8
Price certainty	Under no circumstances must total costs exceed the budget allowance	4	8	10	2	2
Professional responsibility	The project sponsor prefers to retain direct contact with the design team	3	10	1	10	10
Risk avoidance	The project sponsor wishes to pass controllable risk to other parties	3	5	10	5	4
Responsibility	Minimum contractual links preferred	2	5	10	4	1
Total			143	157	133	133

altered to allow the user to rate each of the criteria in terms of the priority that he/she wished them to have. The second deficiency was that the guide assumes that all the listed criteria are of equal importance to the client, but, as each procurement system may have a different degree of relevance to each priority relative to the other procurement paths, it was proposed that a measure of their suitability needed to be built into the process. This was achieved by indicating the relative utility of each procurement path against each criterion on a numerical scale, thus enabling a set of utility factors to be established for use in the decision chart.

These factors were initially determined by means of a method of scoring based upon the work of Fellows and Langford [24], which was confirmed by comparison with a survey of the opinions of five experts. Table 9.5 shows the revised procurement path decision chart, incorporating the various utility factors, which is intended to be completed as follows:

- Having considered all of the client's priority questions, the user enters the relative importance of each criterion in the client's priority rating column on a scale of 1 to 20.
- The rationalised priority rating is then calculated for each criterion by dividing individual priority ratings by the sum of all the ratings and this is entered into the chart. The total of all the rationalised priority ratings should always be equal to 1.
- All of the rationalised priority ratings are then multiplied by each of the utility factors for the various procurement paths and the results entered into the chart.
- The results for each procurement system are then totalled and ranked on the basis that the most appropriate method has the highest total score.

Table 9.6 shows a chart which has been completed for a hypothetical industrial project, where the client requires an industrial unit quickly to realise grant and commence production expeditiously.

The second method reported by Skitmore and Marsden utilised discriminant analysis to examine data collected under a set of criteria that form the characteristics where the various procurement systems are expected to differ. Using these criteria, the researchers were able to discriminate between procurement paths for decision-making purposes.

On the basis of a trial carried out by the authors, this method gave identical answers to those obtained by the use of the multiattribute technique, with both methods producing intuitively correct responses. In theory, the discriminant analysis technique should be more reliable as it uses more of the available information, however it is rather an advanced statistical technique involving a great deal of tedious calculation and probably unsuitable for use by clients or most of their consultants.

Bennett and Grice [25] used Skitmore and Marsden's development of the *Thinking About Building* document as the basis for tabulating the strengths and weaknesses of the various procurement systems so as to provide an opportunity for clients to weight the various criteria in order to reflect their priorities (Table 9.7). The client's priorities and the weighting should, according to the authors, result from a detailed discussion of the issues involved. Before this decision is made, the utility factors allocated to each procurement system should be reviewed in the light of the client's needs and project criteria.

Table 9.8 shows a completed example of the selection method. The researchers point out that the circumstances of the particular client or project could result in the need to pose additional or different questions, or even change the utility factors, before the choice of the procurement system is made.

Such methods of selection that have so far been described are undoubtedly useful in establishing those procurement systems which would not be appropriate and thus identifying the more suitable methods, but, as Bennett and Grice admit, they are based upon a very limited set of criteria. In addition, the procurement systems that are used are not reflective of the very wide, and sophisticated, range of procurement options that are currently available.

The expert computer system package, known as ELSIE [26] and released in 1988 by the Royal Institution of Chartered Surveyors Quantity Surveyors' Division, was intended by its authors to act as the prime mover in setting suitable project management parameters for any building development on the basis of early stage information and, if necessary, before any drawings were prepared.

The package consisted of a report, *The Strategic Planning of Construction Projects* [27], and four linked modules – *budget, procurement, time* and *development appraisal* – held on six floppy discs designed for use on IBM-compatible personal computers. Since its original publication, the system has been taken over and developed by a commercial organisation, which maintains an association with the University of Salford and the RICS and now produces two software programmes ELSIE COMMERCIAL and ELSIE INDUSTRIAL.

The former programme caters for most types of commercial building of up to eighty storeys and will provide a detailed analysis of any internal and external elements of the building. The latter can produce five basic industrial models and is capable of handling 100 three-storey buildings in a single project. Both programmes contain the original four modules:

- *Financial budget* – which enables the user to produce predesign cost estimates and an elemental specification.
- *Procurement* – provides recommendations on the most appropriate procurement route.
- *Time* – provides advice on the likely overall project duration.

Table 9.5 Revised procurement path decision chart, after Skitmore and Marsden [22]

Client's priority questions	Client's priority rating (scale 1–20)	Rationalized priority rating	Procurement paths													
			A Negotiated traditional		B Competitive traditional		C Competitive develop and construct		D Negotiated design and build		E Competitive design and build		F Management contracting		G Turnkey contracting	
			Utility factor	Result	Utility factor	Result	Utility factor	Result	Utility factor	Result	Utility factor	Result	Utility factor	Result	Utility factor	Result
1. Speed How important is early completion to the success of your project?	18	0.3	40		10		60		100		90		110		110	
2. Certainty Do you require a firm price and/or a strict completion date for the project before you can commit yourself to proceed with construction?		0.2	30		30		70		100		100		10		110	
3. Flexibility To what degree do you foresee the need to alter the project in any way once it has begun on site?	4	0.04	110		110		40		40		40		90		10	
4. Quality level What level of quality, aesthetic appearance do you require in the design and workmanship?	10	0.1	110		110		80		40		40		90		20	

Client's priority questions	Client's priority rating (scale 1–20)	Rationalized priority rating	Procurement paths												
			A Negotiated traditional		B Competitive traditional		C Competitive develop and construct		D Negotiated design and build		E Competitive design and build		F Management contracting		G Turnkey contracting	
			Utility factor	Result	Utility factor	Result	Utility factor	Result	Utility factor	Result	Utility factor	Result	Utility factor	Result	Utility factor	Result
5. Complexity Does your building need to be highly specialized, technologically advanced or highly serviced?			100		100		70		50		50		110		20	
6. Risk avoidance and responsibility To what extent do you wish one single organization to be responsible for the project, or to transfer the risks of cost and time slippage?			30		30		70		100		100		10		110	
7. Price competition Is it important for you to choose your construction team by price competition, so increasing the likelihood of a low price?	3		20		110		80		10		80		40		30	
Totals																
Rank order																

Table 9.6 Completed procurement path decision chart for an hypothetical industrial project, after Skitmore and Marsden [22]

Client's priority questions	Client's priority rating (scale 1–20)	Rationalized priority rating	Procurement paths													
			A Negotiated traditional		B Competitive traditional		C Competitive develop and construct		D Negotiated design and build		E Competitive design and build		F Management contracting		G Turnkey contracting	
			Utility factor	Result	Utility factor	Result	Utility factor	Result	Utility factor	Result	Utility factor	Result	Utility factor	Result	Utility factor	Result
1. Speed How important is early completion to the success of your project?	20	0.25	40	10.0	10	2.5	60	15.0	100	25.0	90	22.5	110	27.5	110	27.5
2. Certainty Do you require a firm price and/or a strict completion date for the project before you can commit yourself to proceed with construction?	18	0.22	30	6.6	30	6.6	70	15.4	100	22.0	100	22.0	10	2.2	110	24.2
3. Flexibility To what degree do you foresee the need to alter the project in any way once it has begun on site?	5	0.06	110	6.6	110	6.6	40	2.4	40	2.4	40	2.4	90	5.4	10	0.6
4. Quality level What level of quality, aesthetic appearance do you require in the design and workmanship?	7	0.09	110	9.9	110	9.9	80	7.2	40	3.6	40	3.6	90	8.1	20	1.8

Client's priority questions	Client's priority rating (scale 1–20)	Rationalized priority rating	Procurement paths													
			A Negotiated traditional		B Competitive traditional		C Competitive develop and construct		D Negotiated design and build		E Competitive design and build		F Management contracting		G Turnkey contracting	
			Utility factor	Result	Utility factor	Result	Utility factor	Result	Utility factor	Result	Utility factor	Result	Utility factor	Result	Utility factor	Result
5. Complexity Does your building need to be highly specialized, technologically advanced or highly serviced?	3	0.04	100	4.0	100	4.0	70	2.8	50	2.0	50	2.0	110	4.4	20	0.8
6. Risk avoidance and responsibility To what extent do you wish one single organization to be responsible for the project; or to transfer the risks of cost and time slippage?	17	0.21	30	6.3	30	6.3	70	14.7	100	21.0	100	21.0	10	2.1	110	23.1
7. Price competition Is it important for you to choose your construction team by price competition, so increasing the likelihood of a low price?	10	0.31	20	2.6	110	14.3	80	10.4	10	1.3	80	10.4	40	5.2	30	3.9
Totals Rank order	80	1.00	46.0 7		50.2 6		67.9 4		77.3 3		83.9 1		54.9 5		81.9 2	

Table 9.7 An example of procurement system selection, after Bennett and Grice [25]

Procurement systems

Client's priority:	Traditional				Design and build						Management				Design and manage			
	Sequential		Accelerated		Direct		Competitive		Develop and construct		Management contracting		Construction management		Contractor		Consultant	
Essential 5, 4 / Desirable 3, 2 / Do without 1	Utility	Score	Utility	Score	Utility	Score	Utility	Score	Utility	Score	Utility	Score	Utility	Score	Utility	Score	Utility	Score
Time Is early completion required?	10		50		100		90		60		100		100		90		80	
Cost Is a firm price needed before any commitment to construction is formed?	90		40		100		100		90		20		10		30		20	
Flexibility Are variations necessary after work has begun on site?	100		90		30		30		40		80		90		60		70	
Complexity Is the building highly specialized, technologically advanced or highly serviced?	40		20		20		10		40		100		100		70		80	
Quality Is high quality important?	100		60		40		40		70		90		100		50		60	
Certainty Is completion on time important?	50		30		100		90		70		90		90		100		90	
Is completion within budget important?	30		30		100		100		50		70		60		90		90	
Division of responsibility Is single-point responsibility wanted?	30		30		100		100		70		30		10		90		90	
Is direct professional responsibility wanted?	100		100		10		10		50		70		100		30		30	
Risk Is transfer of responsibility for the consequence of slippages important?	30		30		80		100		70		30		10		100		80	
Results																		

Table 9.8 Completed procurement system selection, after Bennett and Grice [25]

Procurement systems

	Client's priority	Traditional				Direct		Design and build				Management				Design and manage			
		Sequential		Accelerated				Competitive		Develop and construct		Management contracting		Construction management		Contractor		Consultant	
	Essential 5 / 4, Desirable 3 / 2, Do without 1	Utility	Score	Utility	Score	Utility	Score	Utility	Score	Utility	Score	Utility	Score	Utility	Score	Utility	Score	Utility	Score
Time Is early completion required?	2	10	20	50	100	100	200	90	180	60	120	100	200	100	200	90	180	80	160
Cost Is a firm price needed before any commitment to construction is formed?	2	90	180	40	80	100	200	100	200	90	180	20	40	10	20	30	60	20	40
Flexibility Are variations necessary after work has begun on site?	5	100	500	90	450	30	150	30	150	40	200	80	400	90	450	60	300	70	350
Complexity Is the building highly specialized, technologically advanced or highly serviced?	5	40	200	20	100	20	100	10	50	40	200	100	500	100	500	70	350	80	400
Quality Is high quality important?	5	100	500	60	300	40	200	40	200	70	350	90	450	100	500	50	250	60	300
Certainty Is completion on time important?	3	50	150	30	90	100	300	90	270	70	210	90	270	90	270	100	300	90	270
Is completion within budget important?	2	30	60	30	60	100	200	100	200	50	100	70	140	60	120	90	180	90	180
Division of responsibility Is single-point responsibility wanted?	1	30	30	30	30	100	100	100	100	70	70	30	30	10	10	90	90	90	90
Is direct professional responsibility wanted?	3	100	300	100	300	10	30	10	30	50	150	70	210	100	300	30	90	30	90
Risk Is transfer of responsibility for the consequence of slippages important?	3	30	90	30	90	80	240	100	300	70	210	30	90	10	50	100	300	80	240
Results			2030		1600		1720		1680		1790		2330		2400		2100		2120

- *Development appraisal* – provides an evaluation of the feasibility of the project, including the forecasting of finance charges and fees, yields and rental levels and cash flows.

The procurement module is the only element of the ELSIE system that is of concern at this time. The module is able to evaluate the project's characteristics and client priorities and gives guidance on the suitability of five basic procurement systems or their variants – conventional, two-stage conventional, design and build, management contracting and construction management. Once the appropriate programme has been accessed via the computer, a series of questions are posed on the screen which require the user to provide information about the project. The nature of the questions depends on the project being examined and the answers given to previous questions; typical examples of the type of question displayed on the screens are shown in Figures 9.4 and 9.5.

When all of the relevant questions have been answered, the programme evaluates the information and makes recommendations. The guidance is given on screen in the form of a summary containing two elements of information (Figure 9.6):

1　a list of the most appropriate procurement systems ranked in order of suitability; and
2　an indication of the extent to which the various systems will satisfy the client's requirements.

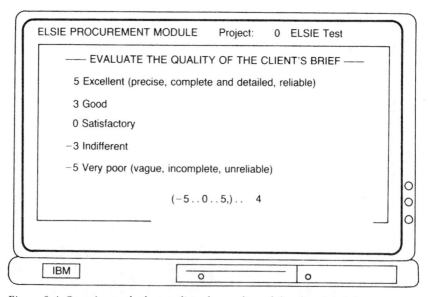

Figure 9.4 Questions asked regarding the quality of the client's brief

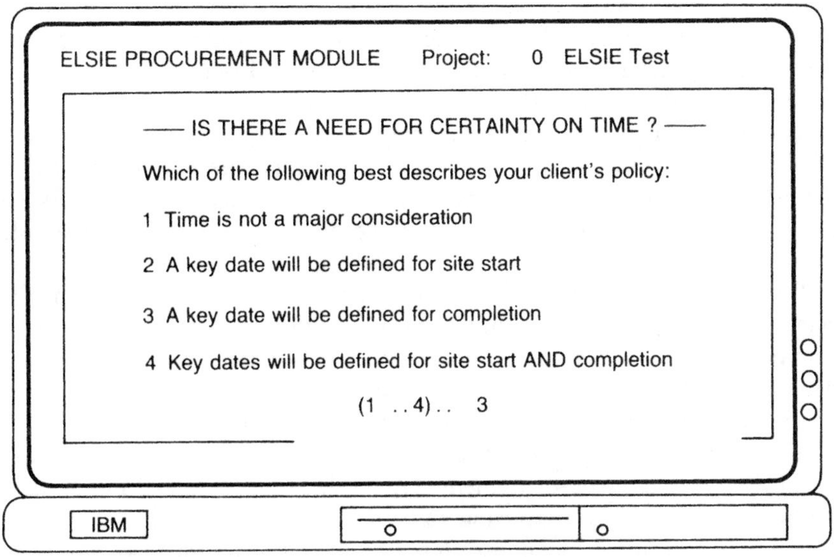

ELSIE PROCUREMENT MODULE   Project: 0   ELSIE Test

—— IS THERE A NEED FOR CERTAINTY ON TIME ? ——

Which of the following best describes your client's policy:

1 Time is not a major consideration

2 A key date will be defined for site start

3 A key date will be defined for completion

4 Key dates will be defined for site start AND completion

(1 ..4).. 3

IBM

Figure 9.5 Questions asked regarding certainty of time

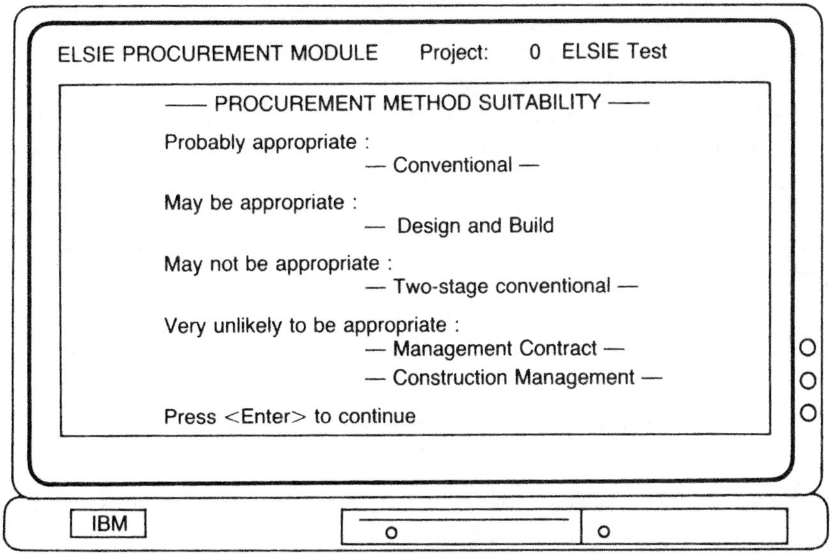

ELSIE PROCUREMENT MODULE   Project: 0   ELSIE Test

—— PROCUREMENT METHOD SUITABILITY ——

Probably appropriate :
— Conventional —

May be appropriate :
— Design and Build

May not be appropriate :
— Two-stage conventional —

Very unlikely to be appropriate :
— Management Contract —
— Construction Management —

Press <Enter> to continue

IBM

Figure 9.6 Procurement method suitability summary

This approach is adopted because, in some situations, the highest rated method may still have a fairly low rating because of the conflicting requirements of the project.

An explanation of the reasons for the guidance that has been given is available in substantial detail via the computer system. The information that has been provided by the user, together with subsequent results, can be stored on a project-by-project basis, and/or printed out so as to obtain a paper record of the process and the recommendations.

The procurement module of the system is therefore yet another method of obtaining advice and guidance on the selection of procurement systems for particular building projects which, it is continually stressed in all the documentation, only provides a second opinion and highlights possible problem areas.

While these guides and aids to procurement system selection point the client, or his advisers, in the right direction, Naphiet and Naphiet [28] suggest that there is no one best method, but rather that what is most appropriate depends on the particular circumstances of the client and his/her project and it follows that the final decision will need to be taken by management taking into account all of the circumstances surrounding the project. The cognitive approach to procurement system selection suggested by Liu [29] supports this view by proposing that 'moderators' such as ability, task complexity and situational constraints should be taken into account during the selection process.

However, it must also be appreciated that it is now generally agreed that a contingency approach to the procurement process is relevant. In other words that there are likely to be, on the same project, many different suitable procurement routes, any, or all, of which will lead to a successful outcome provided that all other aspects of the project strategy have been dealt with correctly and that the project team performs to the appropriate standard.

There is no 'best buy' among procurement systems. Client organisations are complex, and different categories of client require discrete solutions to their procurement needs. Whereas, on a well-planned and -managed project there may well be more than one system that will provide a satisfactory outcome, the choice of system(s) should always be made by matching the client's characteristics, objectives and project criteria to the characteristics of the most appropriate procurement method(s).

9.5 The selection of building procurement systems in practice

Bennett and Grice [25] see the choice of procurement system as a crucial strategic decision, equal only to the establishment of the client's objectives and the decisions taken on the nature of the end product.

An examination of the procurement system selection process reveals that it meets most of the criteria necessary for it to be considered as non-programmable, or unstructured; it is complex, extends over a considerable period of time if carried out correctly, is often constrained by lack of information, is likely to be influenced by a number of individuals and the outcome is often unpredictable.

In addition, the process meets a number of the criteria which differentiate it from lesser decisions, e.g. it is comparatively novel, commits substantial resources, triggers other lesser decisions, and can be organisation-wide in its consequences. Also, the process reflects the characteristics of practical decision-making and choice in general, in that much of it takes place in an environment within which the constraints and consequences of possible actions are not precisely known.

In the past, the way in which clients choose their procurement systems has been found to be unconsidered, automatic and ill-disciplined [30], mirroring the disorderly nature of actual decision-making previously identified. Clients of the construction industry, having taken the primary decision that there is a need to build, will be faced with a choice as to which method of building procurement is the most appropriate in the circumstances. It is this decision that is now examined and, in the light of all of the similarities that have been identified, the actual selection process is examined within the framework used to explore empirical decision-making.

Timing of the selection process

While *Thinking About Building* and other guides to the construction process stress the need to select the procurement route before appointing any individual or organisation, other than a principal adviser, it has been suggested that many clients do not appear to recognise the necessity of making such an early decision or even occasionally realising that such a choice is required.

Figures 9.7 and 9.8 illustrate the timing of the choice of procurement system among sixty-two clients surveyed as part of research carried out by the author in 1994 and show that over 77 per cent of this group chose the procurement system within the inception or feasibility stages. In other words, sufficiently early on in the procurement process as to enable any of the wide range of available procurement systems to be chosen and to ensure that the client has no need to appoint any individual or organisation, other than perhaps an independent adviser, who might be prone to giving subjective advice on the choice. All of the remaining clients who were surveyed made the choice too far into the process – some of them much too far – by leaving the decision until the detailed design was under way and any possibility of an unbiased choice had been removed by the appointment of design consultants.

While a substantial majority therefore appeared to be following recommended practice, the number of organisations that did not make their choice early enough was still too high for any comfort to be drawn from the results, particularly bearing in mind that the majority of the non-conforming organisations were experienced clients.

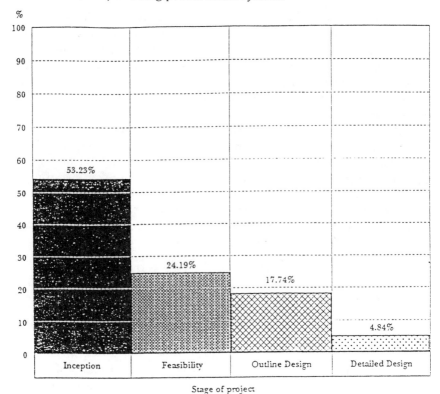

Figure 9.7 Stage of the project at which clients chose their procurement system

Setting objectives

Although there is little doubt that the majority of decisions to build, by their very nature, are made as the result of deliberate managerial strategies formulated within the aims and objectives of the organisation, establishing whether this is the case with the decisions made about the selection of procurement systems is more problematic.

The report of the working party established by government [31] to inquire into the problems of large industrial construction sites found that clients' policies/aims in relation to the implementation of their projects, including the selection of procurement systems, had not in general been designed to meet the unique circumstances of large, expensive, complex and lengthy projects and recommended that 'balanced and compatible policies' needed to be formulated for such projects.

The report *The Public Client and the Construction Industries* [30] incorporated case studies of fifty building and civil engineering projects made up of forty-four from the public sector and six from the private sector. The studies indicated that the surveyed clients were aware of the criteria associated

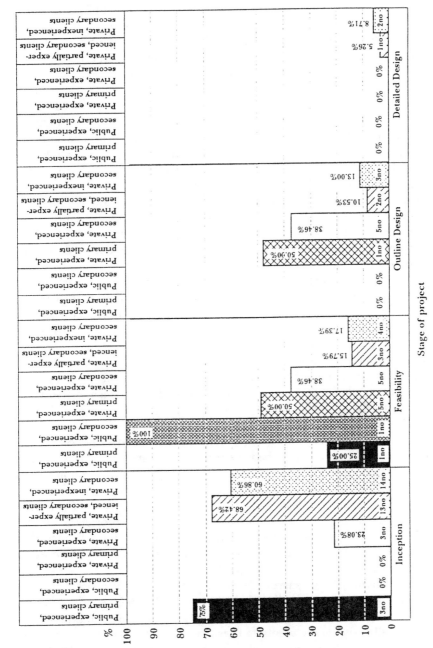

Figure 9.8 Stage of project at which various categories of client chose their procurement system

with successful projects and were able to identify the clients' role as one of defining the scope, objectives and priorities of a project.

Virtually all of the public-sector clients' decisions about the most appropriate method of designing and building their projects were determined by the policy of the organisation, which, in the majority of cases, meant standing orders. In many cases, however, these could have been waived, or amended, so as to use the whole range of procurement systems.

The governmental report *Faster Building for Industry* [32] concluded that, whereas a few of the eighty or so client participants formulated agreed objectives and standards, the majority of those interviewed were vague in formulating their requirements. It would appear that such behaviour did not stem from the organisation's policies, or indeed lack of such guidelines or regulations, but rather that clients were unaware that they had an essential contribution to make to the success of their project.

Cherns and Bryant [33], in a study of clients of the construction industry, suggested that the earliest decisions about the project organisation were determined by the client's organisational culture, procedures and structures and were often idiosyncratic and unduly constrained by residues of the client's preproject history.

A study [28] of six North American and four UK-based projects found that the clients placed great significance on the development of clear and widely understood objectives, and the researchers concluded that it is vital that clients have a clear understanding of their project goals as early in the process as possible.

Hewitt [34], in a survey of twenty-one public and private organisations, found that all of the public-sector clients' and 63 per cent of the private-sector clients' choice of contractual arrangement (procurement system) was affected by the organisation's policies.

In Bresen's study [35] of 138 clients of the construction industry, which was carried out in 1986 and 1987, it was established that 62 per cent of the organisations surveyed classified the procurement system used as 'official policy or normal practice'; once again, this finding was more pronounced in the case of public-sector clients and experienced organisations.

The authors of the 1988 NEDO report *Faster Building for Commerce* [36] carried out detailed case studies on sixty commercial projects and found that experienced customers in particular tended to use a well-tried and trusted method of procurement and that such a system had often been developed to suit the client's own corporate objectives and resources.

The responses to the surveys carried out by the author in 1994 showed that nearly 29 per cent of all the decisions made by all categories of client had been affected by the policies of the organisation. However, experienced and partially experienced clients had been more affected, an average of nearly 42 per cent, than inexperienced clients, an average of 9 per cent. Therefore, it can be reasonably safely concluded that, although a substantial proportion of clients in general are influenced by the policies of their organisation when selecting their building procurement systems, experienced and partially experienced clients are much more affected than inexperienced clients.

While these findings are not unexpected, particularly as inexperienced clients will not have had any previous opportunity to establish policies on procurement, they do highlight the fact that many experienced and partially experienced clients will be restrained by the policies of their organisation when making their choice of procurement system, although not normally to the extent of stipulating the use of a specific system or systems.

Examining the setting of objectives for the delivery of the project itself, whether affected by the client's organisational policies or not, again presents some difficulty because of a lack of the availability of definitive information on the practical implementation of this element of the decision process.

Lindblom's [37] view that there is often difficulty in determining what the 'problem' actually is and/or setting objectives to deal with it is confirmed by the literature related to the formulation of the project brief – the vehicle used for communicating the client's needs and objectives to other participants in the construction process.

A study by MacKinder and Marvin [38] in 1982 indicated that the briefing process in construction projects is usually inadequate, with the necessary information not being forthcoming from the client, or being elicited by his/ her consultants, resulting in an inefficient construction process.

The study *Construction for Industrial Recovery* [39] highlighted the dissatisfaction that many of the 300 or so participants felt about the help given by the industry in developing their project brief, particularly bearing in mind the client's lack of knowledge of the industry and his/her lack of resources to carry out the formulation of the brief.

Bresen *et al.* [35] surveyed 138 building projects and found that in 70 per cent of the case studies only a very basic statement of needs was prepared, after which the brief evolved through consultation and negotiation.

The clients that participated in the *Faster Building for Commerce* [36] study tended to concentrate on the technical aspects of the project and said little about other project objectives at the briefing stage. It was found that the participants did not think through their brief clearly or engage in sufficient dialogue with their consultants.

The setting of project objectives by the client has been examined by a number of authorities. The working party that prepared the Wood Report [30] found that the criteria necessary for a successful project that were consistently mentioned by all of the 300 participants in the case study were: meeting the budget cost, low maintenance costs, time, quality, functionality and aesthetics.

The two government-sponsored studies *Faster Building for Industry* [32] and *Faster Building for Commerce* [36] both examined, as the titles imply, the time aspects of the brief, but the results of the case studies included in both reports suggest that the establishment of this single objective took priority over all other 'management' aspects of the project.

In Chapter 2, the author described the identification of the objectives/ needs of thirty-seven surveyed organisations based upon a standard list of criteria, but in this case relating to the respondent's last, or only, project. While these and other surveyed clients could be considered, on the basis of the evidence, to be aware of the need to establish objectives as one of the initial steps in the selection process, in the case of all of the surveys that have been examined the respondents were provided with a list of criteria to rate, with hindsight, thus leaving open the possibility that in a real situation this disciplined approach might not be adopted.

This possibility appears to be partially supported by the fact that only thirty-seven of the total of sixty-three organisations surveyed by the author responded to the request to rate the listed needs/criteria, with all but one of those that did not participate being private partially experienced or inexperienced secondary clients. Thus, over 41 per cent of all the organisations surveyed were unable, or unwilling, to identify their objectives, even under very controlled circumstances, with this inability being concentrated among the most inexperienced clients.

Finally, it is always necessary to be aware that the temporary multidisciplined organisations that manage procurement systems are characterised by the fact that the client and the other participants in the process have to resolve disparities between the two levels of objectives that are always present in such circumstances, i.e. the specific and short-term project goals and the general and long-term objectives of the organisation to which the participants owe allegiance.

Identifying and developing alternative procurement system solutions

No empirical studies appear to have been carried out into the ways in which clients of the construction industry choose their procurement systems, and it is as a result of this dearth of information that the following sections rely heavily on the author's research. Most researchers have classified the decisions they examined by the process that eventually led to the making of the decision itself.

Nutt [40] identified five process models, i.e. *historical, off-the-shelf, appraisal, search* and *nova*, and carried out the classification of the seventy-three cases he examined by sorting the detailed descriptions of the decision processes into a set of groupings or patterns which were mutually exclusive using the criteria of stability (where within-category similarities and between-category differences can be explicitly specified) and clarity (where the distinctions between categories had theoretical as well as practical significance).

Mintzberg *et al.* [41] found that the decisions they examined fell into seven groups according to the path they took through a basic decision model.

The classification also depended, in large part, on the nature of the dynamic factors that were encountered and the type of solution that was implemented.

Hickson *et al.* [10], on the other hand, identified three ways in which the 136 decisions they examined were made. These were derived from an analysis of twelve process variables which was used to work out a threefold meaningful grouping of cases in which the decisions in each group had more similarities than they had with those in other groups.

While it would have been possible to classify the decisions examined by the author's research in accordance with these predetermined models, it was felt that a better understanding of the decision process leading to the selection of a building procurement system would be achieved if the procedures used by the clients surveyed were examined and classified independently, using the work of the three researchers as a background and basis.

However, as the evaluation procedures that were used by the clients fell more or less spontaneously into groupings, Nutt's method, but not his model, was used to classify the decision processes, with the result that the five distinct categories, or classifications, listed in Table 9.9 were identified.

The five classes had the following characteristics:

- *Analytical search* – where the client's needs were analysed and project criteria were established and matched to the characteristics of the most appropriate procurement system.

  Included in this category were those decision processes where clients also took advice from external consultants, in-house experts and colleagues. This process follows the recommended decision-making route, although no use was made of any of the aids to procurement system selection that were described in previous chapters.
- *Consultative search* – no formal analysis of needs or determination of project criteria were made, but a search of available procurement methods was conducted by a process of discussion with advisers and other experienced client organisations.

  Again, no use was made of aids to procurement system selection.

  A number of variants of this classification were identified which related to the sources of advice; from the organisation's own managers (some of whom had past experience of implementing construction projects), external construction-related consultants, other friendly organisations and building contractors.

  It is always possible that the sources of advice may have themselves carried out some form of analytical search in order to provide the guidance they gave; however, the framing of the questions posed to the clients in the survey questionnaire should have eliminated this possibility.
- *Historical evaluation* – the procedure whereby past experience of the organisation in implementing building projects, or the knowledge of an individual manager of such activities from previous employment, is used

Table 9.9 Classification of procedures used by clients for the evaluation of procurement systems

Procedure adopted for evaluation	Classification of procedure	Total (54)
We analysed our needs, established project criteria and matched these to the characteristics of the most appropriate procurement system	Analytical search	7
The decision was made by analysing our needs, establishing project criteria and matching these to the most appropriate procurement system, we compared the characteristics of all appropriate procurement systems in order to determine which of the various methods were most appropriate, no procedure was used because we instinctively knew that our decision was correct/seemed obvious and company policy determined the choice, we discussed the choice with our professional advisers and our senior management and reached a consensus, we also considered the current state of the construction industry, the availability of experienced consultants and contractors and the reputation of a particular consultant or contractor	Analytical search	1
We analysed our needs, established project criteria and matched these to the most appropriate procurement system and discussed the choice with our professional advisers/senior management and reached a consensus	Analytical search	1
We analysed our needs, established project criteria and matched these to the characteristics of the most appropriate procurement system, discussed selection with other organisations with experience of building their own facilities, took advice from professional advisers, considered the availability of experienced consultants and contractors and recommendations from external sources on specific systems, consultants and contractors	Analytical search	1
We compared the characteristics of all appropriate procurement systems and applied them to our project in order to determine which gave the best results, we also considered the availability of experienced consultants and contractors and the reputation of a particular consultant or contractor	Analytical search	1
We compared the characteristics of all appropriate procurement systems and applied them to our project in order to determine which gave the best results	Analytical search	1
No procedure was used, we discussed the choice with our professional advisers and our own senior management and reached a consensus as to which system to use	Consultative search	5
No procedure was used, our choice was made on the basis of advice obtained from external consultants	Consultative search	4

No procedure was used, we discussed the choice with our professional advisers and senior management and reached a consensus, we also considered the availability of, or need to use, our own resources, our choice was also based upon our past experience or personal knowledge	Consultative search	1
No procedure was used, we consulted other organisations with experience of building their own facilities, we also considered the reputation of a particular consultant or contractor, our choice was also based upon past experience/personal knowledge	Consultative search	1
No procedure was used, we discussed the choice with other organisations with experience of building their own facilities and based our decision on their advice	Consultative search	1
No procedure was used because choice was made on the basis of discussions with other organisations with experience of building their own facilities, advice from contractors and the reputation of a particular contractor	Consultative search	1
No procedure was used because our choice was made on the basis of discussions with our professional advisers and other organisations with experience of building their own facilities	Consultative search	1
No procedure was used because our choice was made on the basis of advice from external consultants and in-house expert and other organisation with experience of building their own facilities	Consultative search	1
No procedure was used, we obtained advice from local contractors	Consultative search	1
No procedure was used as our choice was based upon past experience/personal knowledge	Historical evaluation	15
No procedure was used as our choice was based upon past experience/personal knowledge and the reputation of a particular contractor	Historical evaluation	1
No procedure was used because we instinctively knew our decision was correct/seemed obvious	Intuitive evaluation	4
No procedure was used because we instinctively knew our decision was correct/seemed obvious, our choice was based upon past experience/personal knowledge, we also considered the availability of experienced consultants and contractors and the reputation of a particular consultant or contractor	Intuitive evaluation	1
No procedure was used because company policy determined the choice	Policy compliance	4
No procedure was used because company policy and others determined the choice	Policy compliance	1

to evaluate the alternative solutions and develop the most appropriate of these.

The responses obtained from the surveyed clients show that the majority of the evaluations were based upon the crude proposition that 'we used it before' and that little regard was taken of changes in project criteria, the client's own needs or the current state of the construction market.

- *Intuitive evaluation* – in this routine, there was no search for alternatives or external advice and the identification of alternatives and the development of the eventual solution was based solely on an intuitive evaluation of the problem of selection.

  The responses obtained from the small number of clients whose decisions fell into this classification demonstrated the 'gut-feeling' approach sometimes adopted by mainly inexperienced clients.

- *Policy compliance* – this routine was used where company policy had been established, often as a consequence of financial regulations, to control or restrict the use of procurement systems to those that satisfied the policy.

  Where financial restraints were imposed, for example where only the use of lump sum, fixed price contracts were permitted, some search and development took place among a small number of procurement systems.

Table 9.10 shows how the five classifications of the evaluation processes were distributed among the six categories of client. The table also shows that *analytical search*, i.e. the recommended theoretical method, was carried out by only 22 per cent of all the respondents; *consultative search* and *historical evaluation* by 30 per cent each; and *intuitive evaluation* and *policy compliance* both accounted for 9 per cent. Thus, less than one-quarter of the surveyed clients used the method of identifying and evaluating the various procurement systems that was best suited to their needs and the requirements of the project.

Looking at the behaviour of individual categories of client, it will be seen that although the number of cases examined in the first three client categories was too small to allow valid conclusions to be drawn the use of the analytical search routine was minimal:

- Public/experienced/primary clients used three routines, i.e. *analytical search*, *historical evaluation* and *policy compliance*.
- The single public/experienced secondary client used *historical evaluation*.
- Both of the private/experienced primary clients also used *historical evaluation*.

Responses that were obtained from larger numbers of the other clients showed that:

Table 9.10 Incidence of the use of individual evaluation process classifications by various categories of clients

	Number/percentage of responses							
Classification of evaluation process	Public experienced primary clients (4)	Public experienced secondary clients (1)	Private experienced primary clients (2)	Private experienced secondary clients (13)	Private partially experienced secondary clients (13)	Private inexperienced secondary clients (21)	Total (54)	
Analytical search	1			3	4	4	12	a
	1.85			5.55	7.41	7.41	22.22	b
	25.00			23.08	30.77	19.05	N/A	c
Consultative search				3	3	10	16	a
				5.55	5.55	18.53	29.63	b
				23.08	23.08	47.62	N/A	c
Historical evaluation	1	1	2	5	5	2	16	a
	1.85	1.85	3.705	9.26	9.26	3.705	29.63	b
	25.00	100.00	100.00	38.46	38.46	9.52	N/A	c
Intuitive evaluation					1	4	5	a
					1.85	7.41	9.26	b
					7.69	19.05	N/A	c
Policy compliance	2			2		1	5	a
	3.705			3.705		1.85	9.26	b
	50.00			15.38		4.76	N/A	c

Notes
a Number.
b Percentage of total.
c Percentage of number in client category.

- Private/experienced/secondary clients used four of the five classes of process, *analytical search, consultative search, historical evaluation* and *policy compliance.*
- Private/partially experienced/secondary clients also used four of the routines, *analytical search, consultative search, historical evaluation* and *intuitive evaluation.*
- Private, inexperienced/secondary clients used all of the routines.

The performance of experienced and partially experienced private clients was similar in a number of ways, with both categories carrying out consultative search and historical evaluation at, respectively, approximately 23 per cent and 38 per cent. Twenty-three per cent of the experienced clients and 31 per cent of the partially experienced respondents used analytical search.

Only 20 per cent of the private inexperienced clients carried out an analytical search, with the majority of the group opting for the use of consultative search, although nearly 20 per cent relied on intuitive evaluation.

What conclusions can be drawn from these results?

Although the numbers were small, and any conclusions must be heavily conditioned by this, public and private experienced clients appear to carry out this element of the procurement selection process mainly by means of historical evaluation and, in the case of primary clients, are constrained by policy compliance. If this behaviour is typical of these categories of client, it would appear to indicate a surprising lack of understanding of the principles of procurement selection, resulting in a heavy reliance on possibly inappropriate past practice and compliance with policies and regulations that could be detrimental to the interests of the organisation.

Although the private experienced and partially experienced secondary respondents carried out an analytical search and a consultative search for a large proportion of their projects, nearly 40 per cent of both groups relied on the results of what may have been inappropriate past projects upon which to base their assessments.

Finally, private inexperienced secondary clients behaved much as expected by relying on consultative search for nearly half of their evaluations, with a worrying one-fifth of the respondents relying on their intuition!

So, once again, the results indicate that clients do not in general terms behave in a logical, disciplined, rational way when proceeding through this part of the decision-making process.

Evaluation and selection of alternatives and final choice

While a number of elements – for example evaluation of benefits, costs of alternatives and estimation of risk (as to whether alternatives will fulfil objectives) – should be present during this phase, it is not always possible in reality to isolate these; the examination of this element of the process is made in the context of that reservation.

Before considering any of these constituents, it should be noted that research (see Table 9.11) has shown that although many of the public and private experienced clients were knowledgeable about virtually all of the eight main procurement systems most of the partially experienced and inexperienced clients were unaware of many of the methods that were available to them.

Although this result would not in itself be surprising, the fact that most of the less experienced clients obtained advice from construction professionals obviously provided no assurance that all possible solutions were considered when choices were made. The author's research has shown that many consultants' knowledge of the characteristics of the various procurement systems is incomplete. It is therefore apparent that, in the case of the less experienced organisations, the selection of procurement systems was made without knowledge of the characteristics, advantages and disadvantages of all but a small number of the wide range of procurement systems, as well as an insufficient appreciation of the workings of the construction industry.

It was not possible to establish from the responses to the survey whether or not clients had evaluated the non-financial aspects of all the available alternatives, although a number of indications emerged from such information as had been provided.

First, as the figures in Table 9.9 show, only 20 per cent of all of the surveyed organisations followed the analytical search routine during the evaluation process, with 30 per cent relying on consultative search. In the case of the former, the process demands that the characteristics of the various alternatives are compared with the client's needs and project criteria, thus ensuring that a comprehensive evaluation takes place. In the case of the latter, it is safe to assume that some discussion and evaluation must have taken place during the consultation process or have been carried out by the consultants.

It follows that a maximum of half of the clients that were examined could have assessed the non-financial benefits of the various alternatives; the remainder, because of the evaluation routines that they followed, would not have carried out this function.

Second, it has been established that some 40 per cent of all of the clients surveyed were unable, or unwilling, to identify the objectives/needs which they applied to their project, indicating that they had not carried out an evaluation of alternatives during the decision-making process – the majority (96 per cent) of these were partially experienced or inexperienced clients.

There was no evidence to determine whether any evaluation of the financial consequences of using the identified alternatives was carried out by any of the clients, but, as little guidance exists as to the real cost of using different procurement systems [42], it would be surprising if such evaluations were carried out even by those clients who adopted the analytical search approach to selection.

Considering the prominent place held by the need to 'obtain value for

Table 9.11 Number of respondents having knowledge of the main procurement systems

	Number of respondents						
Procurement system	Public experienced primary clients (4)	Public experienced secondary clients (1)	Private experienced primary clients (2)	Private experienced secondary clients (13)	Private partially experienced secondary clients (11)	Private inexperienced secondary clients (16)	Total (47)
Conventional	4	1	2	13	11	14	45
Design and build	4	1	2	13	11	9	40
Package deal	3	1	2	9	6	6	27
Turnkey	3	1	2	9	8	5	28
Design and manage	3	1	2	11	6	6	29
Management contracting	3	1	2	12	3	3	24
Construction management	3	1	2	7	5	2	20
British Property Federation system	2	1	2	3	1	1	9
No knowledge of any system						2	2

money' among the surveyed clients' objectives, one must ask how these clients could be satisfied that the procurement system they had selected using a non-analytical routine would meet this criterion?

There was no evidence that the surveyed clients had made any attempt to assess formally the likelihood of the various alternatives meeting the objectives laid down at the briefing stage of the project, although it could be argued that those organisations that carried out an analytical search or even a consultative search did, as part of these routines, examine the consequences of the alternative choices.

A case could be, and no doubt would be, made that those who took the historical evaluation and policy compliance routes had based their decisions on experience and organisational regulations that had in the past achieved the objectives of the organisation, but objectives change and decisions that were taken in the past may not be relevant today.

Hickson *et al.* [10] found that the interests, i.e. the parties having influence on the various parts of the decision process, affected the way in which the alternative solutions were examined. In the selection processes that the author examined, both internal and external interests were involved and appeared to influence the routines in the following way. Half of the selection routines implemented by experienced clients were affected by company policy, but only one of the total number involved an in-house 'expert' in the process; no other external interests were involved.

The inexperienced and partially experienced clients surveyed involved more interested parties in their selection processes, with in-house 'experts' in these cases being consulted fourteen times (21.88 per cent) out of the total of sixty-four consultations. These two groups of clients also took advice from friendly external organisations, professional consultants, contractors and others.

Table 9.12 shows the consultations, etc. that were made by various categories of client.

The evidence shows that experienced clients' decisions are often affected by the policies of their organisation but that few other interests are taken into account during the process, probably as a result of the guidance provided by the policies and their own expertise in construction matters. On the other hand, inexperienced and partially experienced clients, as might be expected, are influenced by more internal and external interests, particularly professional consultants and in-house experts, presumably as a result of their lack of expertise and knowledge and their need to obtain substantial amounts of advice, reassurance and 'approval'.

Turning to the question of final choice, and commitment to a specific alternative, it was possible to determine in more detail how each of the surveyed clients selected the procurement system they used. An examination of the responses received from sixty-one of the surveyed clients enabled three distinct methods of choice to be identified using Nutt's method of classification:

Table 9.12 Internal and external consultations made by clients during the selection process

	Number of consultations, etc.						
Interests, etc. affecting selection process	Public experienced primary clients (4)	Public experienced secondary clients (1)	Private experienced primary clients (2)	Private experienced secondary clients (13)	Private partially experienced secondary clients (13)	Private inexperienced secondary clients (21)	Total (54)
Company policy	2		2	5	5	2	17
Consultations in-house 'expert'		1		1	12	2	15
Consultants					16	24	40
Contractor					2	5	7
Friendly external organisation					1	2	3
Own directors					1	1	2
Funder/grant aider					1		1
Development agency						1	1

- *Fulfilment choice* – where the needs of the client and/or the project criteria were seen to be fulfilled by the chosen procurement system.

   This routine included complying with company policy, satisfying the need for single-point responsibility and the desire to use the organisation's own in-house designers.

- *Advisory choice* – where the choice was made on the advice of external consultants and internal experts or experienced employees, contractors and other friendly experienced organisations.

   The reliance on in-house experts and experienced employees was substantial, with nearly 22 per cent of the consultations being held with this source of advice by partially experienced and inexperienced clients. No information was available as to the level of expertise of the advisers that were consulted but it is likely that the advice obtained would be based on past experience rather than any specialist expertise, and some of the decisions in question might therefore have been placed in the next category.

- *Historical choice* – where the choice of procurement method was made mainly, or solely, on the basis of previous satisfactory experience of a specific system.

   The use of this routine was coupled, occasionally, with taking advice from external and internal 'consultants', reinforcing the view that there may well be some overlap between this method and the advisory choice category.

Table 9.13 shows the numbers of the three routines used by the six client categories.

It was observed that private/partially experienced/secondary clients used the historical choice route to make the majority of their decisions, whereas private/inexperienced/secondary clients favoured the use of the advisory choice routine. This reinforces the commonsense-based, but tentative, conclusions drawn from other elements of the study that inexperienced clients make decisions on the basis of expert advice, or even by delegating the decision itself to such experts, while experienced and partially experienced clients tend to base their choices on experience.

Apart from the difference in the number of responses between these results and those relating to the classification of evaluation processes in Table 9.8, there was evidence of some discrepancy between the classification of the responses in the way the process was carried out and the method used to make the actual selection.

These discrepancies appear to stem from the fact that although the evaluation process routine followed, for example, company policy, the actual selection was made on the basis of satisfying the project criteria, which were of course based upon company policy! The number of discrepancies was small in relation to the overall number of choices and did not invalidate any conclusions that may be drawn from the classification exercises.

Table 9.13 Method used by various categories of client to choose their building procurement systems

Method of choice	Public experienced primary clients		Public experienced secondary clients		Private experienced primary clients		Private experienced secondary clients		Private secondary partially experienced clients		Private inexperienced secondary clients		Total	
	No.	%	No.	%	No.	%	No.	%	No.	%	No.	%	No.	%
Fulfilment choice	2	75.00			1	50.00	5	38.46	3	17.65	6	25.00	18	29.51
Advisory choice	1	25.00	1	100.00			2	15.38	4	23.53	16	66.67	24	39.34
Historical choice					1	50.00	6	46.16	10	58.82	2	8.33	19	31.15

Despite this caveat, it was found that, when comparing the evaluation and choice processes, analytical search and policy compliance led to fulfilment choice; consultative search and intuitive evaluation culminate in advisory choice; and historical evaluation results in the use of historical choice.

The large number of consultations carried out by the surveyed inexperienced clients and their lack of knowledge of the full range of procurement systems, and of the construction industry itself, confirm Pettigrew's [43] findings that one of the major causes of uncertainty in strategic decision-making is the complex and technical environment in which decisions are made.

There was evidence, particularly among the less experienced clients, of a further critical source of uncertainty in that the criteria used in the evaluation were ill-defined, or not even identified. An examination of the results of the survey shows that only just over 40 per cent of the partially experienced and inexperienced clients were able to identify their project criteria despite the fact that they were presented with a list of such criteria from which to make a choice.

Where the historical choice routine was used, uncertainty and risk appear to have been ignored, presumably on the grounds that previous experience meant that such difficulties could be avoided as they had been in the past. This approach gives more credibility to such experience than is perhaps warranted, and supports Bass's [44] views that such an approach is a common, if somewhat suspect, corporate technique.

On the question of risk, in the context of selection, it has been established that uncertainty is present at both the evaluation and choice stage as a consequence of a lack of information about future events. A number of the categories of client rated the elimination and/or reduction of risk as being of importance when selecting procurement systems, but there was little evidence in any of the responses of any attempt to consider seriously this aspect of the choice process.

On the question of authorisation of choice, it was apparent that, where firm policy guidelines had been established, managers had made choices that reflect such guidance without reference back to any senior director or board member. There is also the implication that the views of consultants are often considered to be so valid as to be accepted and seen by managers as virtual authorisation of the use of any alternative that the managers themselves may have chosen with, or without, the involvement of senior management.

However, among the responses received from the surveyed clients of all categories, there was no evidence that authorisation of the final choice had caused any major difficulties or was even considered to be a separate element of the selection process.

Implement decision and monitor success

The surveyed clients were not asked how they implemented their decisions, although it can safely be assumed that the majority used one of the standard procedures for the appointment of consultants and contractors available within the construction industry.

All clients were asked to confirm whether they were dissatisfied, reasonably satisfied or completely satisfied with the procurement system they had chosen. No experienced clients, only a comparatively small percentage of partially experienced (5.26 per cent) and a larger number of inexperienced clients (12.5 per cent) expressed dissatisfaction.

A greater number of clients from five categories were only reasonably satisfied:

- public/experienced/primary – 25.00 per cent;
- public/experienced/secondary – 100.00 per cent;
- private/experienced/secondary – 23.08 per cent;
- private/partially experienced/secondary – 31.58 per cent;
- private/inexperienced/secondary – 25.00 per cent.

For all sixty-three of the surveyed clients, the overall level of satisfaction was:

- dissatisfied – 6.35 per cent;
- reasonably satisfied – 26.98 per cent;
- completely satisfied – 66.67 per cent.

A number of reasons for this lack of dissatisfaction could be suggested. In the case of the four categories of experienced clients, it could be that they were totally satisfied as a result of the active involvement in their projects that is a characteristic of such clients, who consider expertise in project management to be a necessary part of their in-house capabilities. In the case of partially experienced and inexperienced clients, it is suggested that the high level of satisfaction may be due to the lack of knowledge as to what the industry can achieve when an efficient project team uses the correct procurement system.

Rowlinson and Newcombe [45] found that, in studying clients' perceptions of the success of a project, although a link between achieving cost and time targets and satisfaction existed a significant number of clients still expressed satisfaction when either, or both, were exceeded.

It was suggested that this phenomenon might be explained by the fact that experienced clients allow a contingency in their internal budgets for the construction phase and/or that the client, having built up a rapport with the building team and being informed regularly and accurately of the time and cost implications throughout the project, will not be dissatisfied if overruns occur.

In an attempt to establish whether these theories were correct, or whether other reasons could be identified, the levels of satisfaction of the clients were compared with their knowledge of the available procurement systems. Table 9.14 illustrates these comparisons for experienced clients.

It was observed that – of twenty experienced clients – five were aware of only four, or fewer, of the main procurement systems. As four out of these five clients were completely satisfied with the outcome of their project, it could be argued that their satisfaction was misplaced and that their standards were lower, in terms of acceptable performance, than if they had been sufficiently knowledgeable. It is possible that this state may result in clients accepting a lower standard of performance and satisfaction than is really necessary. Table 9.15 illustrates similar comparisons relating to partially experienced and inexperienced clients.

The phenomenon of organisations with knowledge of a very small number of procurement systems being either reasonably or completely satisfied was still present, with 55 per cent of partially experienced clients and 69 per cent of inexperienced clients having knowledge of four or fewer procurement methods.

In the case of the partially experienced respondents, only one was dissatisfied, one was reasonably satisfied and four were completely satisfied. Of the inexperienced clients, three were dissatisfied, two reasonably satisfied and six completely satisfied.

Although the same argument could be made as for experienced clients, i.e. that their lack of knowledge meant that their satisfaction was misplaced, it could also be argued that a substantial percentage of partially experienced and inexperienced clients are not in possession of all of the requisite information about procurement systems to enable them to select the most appropriate method and that this may result in them achieving a less satisfactory outcome to their project than necessary.

In a further effort to determine whether the satisfaction expressed by the majority of surveyed clients was misplaced, all of the projects in which respondents had rated their needs/criteria were assessed by means of the multiattribute selection technique formulated by Skitmore and Marsden [22] to establish which procurement system should theoretically have been used. Table 9.16 summarises the results of this exercise.

The comparison showed that the system chosen by the clients coincided with that selected by using Skitmore and Marsden's method on six out of eighteen occasions, or, in simplistic terms, the clients' choices were 'wrong' 67 per cent of the time. However, these results need to be heavily conditioned:

- The relative importance of each of the criterion upon which the client's priority rating is based was determined solely by using the responses obtained from the questionnaires completed by the surveyed clients.
- There was no opportunity to discuss the projects in any detail with the clients and a number of instances were observed where the rating of the criteria appeared to be in contradiction to the nature of the project and/ or other responses.

Table 9.14 Comparison of level of satisfaction and knowledge of procurement systems – experienced clients

Category (reference of client)		Level of satisfaction	Number of procurement systems known	Procurement systems that were not known to the client
Public/experienced/primary	(1)	Reasonably satisfied	8	–
Public/experienced/secondary	(4)	Reasonably satisfied	7	BPF
Private/experienced/secondary	(8)	Reasonably satisfied	4	T/key; D and M; C/man; BPF
Private/experienced/secondary	(10)	Reasonably satisfied	7	BPF
Private/experienced/secondary	(17)	Reasonably satisfied	7	BPF
Public/experienced/primary	(2)	Completely satisfied	3	PK/deal; T/key; D and M; M/con; C/Man
Public/experienced/primary	(3)	Completely satisfied	7	BPF
Public/experienced/primary	(20)	Completely satisfied	7	BPF
Private/experienced/primary	(6)	Completely satisfied	8	–
Private/experienced/primary	(7)	Completely satisfied	8	–
Private/experienced/secondary	(5)	Completely satisfied	6	C/man; BPF
Private/experienced/secondary	(9)	Completely satisfied	5	PK/deal; C/man; BPF
Private/experienced/secondary	(11)	Completely satisfied	8	–
Private/experienced/secondary	(12)	Completely satisfied	4	PK/deal; M/con; C/man; BPF
Private/experienced/secondary	(13)	Completely satisfied	6	C/man; BPF
Private/experienced/secondary	(14)	Completely satisfied	8	–
Private/experienced/secondary	(15)	Completely satisfied	7	T/key
Private/experienced/secondary	(16)	Completely satisfied	4	PK/deal; T/key; D and M
Private/experienced/secondary	(18)	Completely satisfied	4	BPF
Private/experienced/secondary	(19)	Completely satisfied	7	PK/deal; T/key; C/man; BPF

Note
BPF, British Property Federation; T/key, turnkey; D and M, design and manage; C/man, construction management; PK/deal, package deal; M/con, management contracting.

- The majority of the projects, while ranging in value from £85,000 to £6.5m, were not complex or likely to require the use of any of the more sophisticated procurement systems; as a consequence, many of the systems chosen when using Skitmore and Marsden's method were only rated marginally higher than the next lowest system, which in some cases had been the client's choice.
- In eight out of the twelve decisions where a different choice had been made from that selected by the client, complete satisfaction had been expressed by the user. If the contingency theory of procurement system selection is accepted, the high level of satisfaction may stem from the very effective management of all other aspects of the project, resulting in the 'incorrect' choice of procurement system becoming relatively unimportant.
- The relatively small sample of clients examined inhibits the drawing of any firm conclusions.

Not withstanding these caveats, there is sufficient evidence to show that two-thirds of the surveyed clients did not choose the procurement system that best matched their stated project criteria. Such results suggest that an incorrect approach to procurement system selection is being adopted by a substantial number of clients.

So what, in summary, are the characteristics of the real selection process?

9.6 Summary

A significant proportion of the surveyed clients left the matter of the choice of the most appropriate procurement system until a stage in the process when any possibility of making an unbiased decision had been removed.

While clients in general are affected by the policies of their organisations when selecting building procurement systems, experienced and partially experienced clients are affected more than inexperienced clients. Although 59 per cent of the surveyed organisations identified and ranked their project objectives and needs, the remaining 41 per cent were unable, or unwilling, to carry out this task even when presented with a list of criteria. This inability suggests that a substantial proportion of clients did not adopt a disciplined approach to this activity and thus were unable or unwilling to establish essential elements of the project brief.

Five distinct categories of the procurement system development and evaluation process were found among the surveyed clients, i.e. analytical search, consultative search, historical evaluation, intuitive evaluation and policy compliance.

Fewer than one-quarter of the processes were carried out by analytical search, the prescribed theoretical method, and none of the clients used any of the numerous aids and guides to building procurement system selection.

Although experienced clients' choices are often affected by the policies of

Table 9.15 Comparison of level of satisfaction and knowledge of procurement systems – partially experienced and inexperienced clients

Category (reference of client)		Level of satisfaction	Number of procurement systems known	Procurement systems that were not known to the client
Private/partially experienced/ secondary	(17)	Dissatisfied	2	PK/deal; T/key; D and M; M/con; C/man; BPF
Private/partially experienced/ secondary	(12)	Reasonably satisfied	7	BPF
Private/partially experienced/ secondary	(33)	Reasonably satisfied	5	D and M; M/con; BPF
Private/partially experienced/ secondary	(34)	Reasonably satisfied	7	BPF
Private/partially experienced/ secondary	(38)	Reasonably satisfied	3	PK/deal; T/key; M/con; C/man; BPF
Private/partially experienced/ secondary	(39)	Reasonably satisfied	8	–
Private/partially experienced/ secondary	(2)	Completely satisfied	3	PK/deal; D and M; M/con; C/man; BPF
Private/partially experienced/ secondary	(28)	Completely satisfied	3	PK/deal; D and M; M/con; C/man; BPF
Private/partially experienced/ secondary	(38)	Completely satisfied	3	PK/deal; T/key; M/con; C/man; BPF
Private/partially experienced/ secondary	(41)	Completely satisfied	3	T/key; D and M; M/con; C/man; BPF
Private/partially experienced/ secondary	(47)	Completely satisfied	5	M/con; C/man; BPF
Private/inexperienced/secondary	(21)	Dissatisfied	0	None known
Private/inexperienced/secondary	(23)	Dissatisfied	1	D and B; PK/deal; T/key; D and M; M/con; C/man; BPF

Category	Satisfaction		Procurement methods
Private/inexperienced/secondary (40)	Reasonably satisfied	3	T/key; D and M; M/con; C/man; BPF
Private/inexperienced/secondary (7)	Reasonably satisfied	1	D and B; PK/deal; T/key; D and M; M/con; C/man; BPF
Private/inexperienced/secondary (10)	Reasonably satisfied	1	D and B; P/deal; T/key; D and M; M/con; C/man; BPF
Private/inexperienced/secondary (1)	Completely satisfied	5	M/con; C/man; BPF
Private/inexperienced/secondary (3)	Completely satisfied	4	T/key; M/con; C/man; BPF
Private/inexperienced/secondary (6)	Completely satisfied	3	PK/deal; T/key; M/con; C/man; BPF
Private/inexperienced/secondary (9)	Completely satisfied	8	–
Private/inexperienced/secondary (11)	Completely satisfied	0	None known
Private/inexperienced/secondary (13)	Completely satisfied	6	C/man; BPF
Private/inexperienced/secondary (16)	Completely satisfied	2	PK/deal; T/key; D and M; M/con; C/man; BPF
Private/inexperienced/secondary (29)	Completely satisfied	5	M/con; C/man; BPF
Private/inexperienced/secondary (32)	Completely satisfied	5	PK/deal; D and M; BPF
Private/inexperienced/secondary (42)	Completely satisfied	1	D and B; PK/deal; T/key; D and M; M/con; C/man; BPF
Private/inexperienced/secondary (44)	Completely satisfied	1	D and B; PK/deal; T/key; D and M; M/con; C/man; BPF

Note
BPF, British Property Federation; C/man, construction management; D and B, design and build; D and M, design and manage; M/con, management contracting; T/key, turnkey; PK/deal, package deal.

Table 9.16 Comparison of client choices with those made by using Skitmore and Marsden's [22] selection system

Category of client	Level of satisfaction	Number of procurement systems known	System chosen by client	System chosen using Skitmore and Marsden's method	Result Same choice	Result Different choice	Project details
Private/ inexperienced/ secondary	Completely satisfied	5	Design and manage	Design and build		x	Bespoke furniture manufacturing factory £4.5m
Secondary	Completely satisfied	3	Conventional	Management contracting		x	Conversion warehouse to paper-making m/room £460k
Secondary	Completely satisfied	8	Conventional	Conventional	x		Hospice building £1.3m
Secondary	Reasonably satisfied	1	Conventional	Design and build		x	Laboratory expansion £150k
Secondary	Completely satisfied	6	Conventional	Design and build		x	Bespoke engine manufacturing factory and offices £3.9m
Secondary	Completely satisfied	1?	Design and build	Design and build	x		Bespoke refrigerator manufacturing factory and offices £260k
Secondary	Reasonably satisfied	1?	Design and manage	Design and build		x	Bespoke wire harness assembly factory £?
Secondary	Completely satisfied	2?	Design and build	Design and build	x		Bespoke warehouse and bakery £300k

Secondary	Completely satisfied	1	Conventional		Design and build	x	Demolition of existing and construct new church £493k
Private/partially experienced/secondary	Completely satisfied	3	Conventional		Design and build	x	Bespoke injection moulding factory £2.5m
Secondary	Completely satisfied	2?	Design and build	x	Design and build		Warehouse and office extension £281k
Secondary	Reasonably satisfied	7	Design and build	x	Design and build		Bespoke hygiene product factory £2.0m
Secondary	Reasonably satisfied	3?	Design and manage		Design and build	x	Warehouse £85k
Secondary	Completely satisfied	6?	Conventional		Design and build	x	Project hall laboratories and offices £450k
Secondary	Completely satisfied	4?	Management contracting		Design and build	x	Car showroom and workshop £350k
Secondary	Reasonably satisfied	3	Design and manage		Design and build	x	Bespoke amenity blocks £210k
Secondary	Completely satisfied	3	Conventional	x	Conventional		Sheltered housing and ancillary facilities £6.5m
Secondary	Completely satisfied	3?	Conventional		Design and build	x	Girls' boarding house £650k
				Total		6	12

Note
?, uncertain information.

their organisations, little account is taken of internal or external influences during the decision-making process. Partially experienced and inexperienced clients on the other hand are much less affected by organisational policies but become involved with many more internal and external interested parties.

Three distinct methods of choice, i.e. fulfilment, advisory and historical, were identified among the surveyed clients and it was found that their use was more or less equally distributed. Specifically, experienced clients favoured fulfilment and historical choice, partially experienced clients opted in the main for historical choice and inexperienced clients mainly used the advisory choice routine.

Although only just over 6 per cent of all of the surveyed clients expressed dissatisfaction with the procurement systems they had chosen, a comparison of the method they had used with the system that they should have selected revealed that two-thirds of the choices that were examined were to some degree inappropriate. Despite this finding being heavily conditioned, it is concluded that a substantial proportion of all clients are likely to be adopting an incorrect approach to the selection of procurement systems and that their satisfaction is thus very often unknowingly based upon the acceptance of lower levels of success than are really necessary.

In general, the evidence from these surveys of clients' decision-making, relative to the selection of building procurement systems, confirms conclusions reached by research in the field of general practical decision-making that: the process often commences with a lack of understanding of the decision situation, minimum knowledge of alternative solutions and very little idea of how to evaluate them and make the final choice.

References

1 Fowler, H.W. and Fowler, F.G. (eds) (1964) *The Concise Oxford Dictionary of Current English,* 5th edn, Oxford: Clarendon Press.
2 Harrison E.F. (1975) *The Managerial Decision-making Process,* Boston: Houghton Mifflin Company.
3 Simon H.A. (1983) *Reason in Human Affairs,* Stanford, CA: Stanford University Press.
4 March, J.G. and Simon, H.A. (1958) *Organisations,* New York: Wiley.
5 Cyert, R.M. and March, J.G. (1963) *A Behavioural Theory of the Firm,* New York: Englewood Cliffs.
6 Allison, G.T. (1972) *Essence of Decision: Explaining the Cuban Missile Crisis,* Boston: Little, Brown.
7 Thomas, H. (1972) *Decision Theory and the Manager,* London: Pitman Publishing.
8 Hodge, B.J. and Johnson, H.J. (1970) *Management and Organisational Behaviour,* New York: Wiley.
9 Bass, B.M. (1983) *Organisational Decision-making,* New York: Irwin.
10 Hickson, D.J., Butler, R.J., Cray, D., Mallory, G.R. and Wilson, D.C. (1986) *Top Decisions: Strategic Decision-making in Organisations,* San Francisco, CA: Jossey-Bass.

11 Simon, H.A. (1960) *The New Science of Management Decision*, New York: Harper and Row.
12 Maccrimmon, K.R. and Taylor, R.N. (1976) 'Decision making and problem solving', in *Handbook of Industrial and Organisational Psychology*, Dunnette, M.D. (ed.), New York: Wiley.
13 DIO International Research Team (1983) 'A contingency model of participative decision-making: an analysis of 56 decisions in three Dutch organisations', *Journal of Occupational Psychology* 56, 1–18.
14 Cole, G.A. (1990) *Management – Theory and Practice*, London: DP Publishing.
15 Herman, S.M. (1978) 'Organisational development', in *Behavioral Science Interventions for Organisational Improvement*, French, W.L. and Bell, C.H. (eds), Englewood Cliffs, NJ: Prentice Hall.
16 Goodacre, P.E. (1982) 'A clients' guide', occasional paper no. 1, Department of Construction Management, University of Reading.
17 Construction Round Table (1995) *Thinking About Building*, London: The Business Round Table.
18 HM Treasury, Public Competition and Purchasing Unit (1991) *Guidance Document No. 33: Project Sponsorship*, London: HMSO.
19 HM Treasury, Central Unit on Purchasing (1992) *Guidance Document No. 36: Contract Strategy Selection for Major Projects*, London: HMSO.
20 Mohsini, R. and Davidson, C.H. (1989) 'Building procurement – key to improved performance', paper to Chartered Institute of Building International Workshop on Contractual Procedures, Liverpool.
21 Birrell, G.S. (1992) 'Choosing between building procurement approaches', in *Concepts and Decision Factors in Management, Quality and Economics in Building*, Bezelga, A. and Brandon, P. (eds), London: E & FN Spon.
22 Skitmore, R.M. and Marsden, D.E. (1988) 'Which procurement system? Towards a universal procurement selection technique', *Construction Management and Economics* 6, 71–89.
23 Franks, J. (1990) *Building Procurement Systems. A Guide to Building Project Management*, Ascot: Chartered Institute of Building.
24 Fellows, R.F. and Langford, D.A. (1980) *Decision Theory and Tendering, Building Technology and Management*, Ascot: Chartered Institute of Building.
25 Bennett, J. and Grice, A. (1990) 'Procurement systems for building', in *Quantity Surveying Techniques*, Brandon, P.S. (ed.), London: BSP Professional Books.
26 Ashworth, G. (1988) 'ELSIE, the QS's thinking friend', *Building* 17 June, 97.
27 Brandon, P.S., Basdon, A. and Hamilton, I.W. (1988) *Expert Systems: The Strategic Planning of Construction Projects*, London: Royal Institution of Chartered Surveyors.
28 Naphiet, H. and Naphiet, J. (1985) *The Management of Construction Projects: Case Studies from the USA and UK*, Ascot: Chartered Institute of Building.
29 Liu, A.N.N. (1994) 'From act to outcome – a cognitive model of construction procurement', in *Proceedings of CIB W-92 International Procurement Symposium, East Meets West*, University of Hong Kong, 4–7 December, Hong Kong: Commission for International Building.
30 Building Economic Development Committee (1975) *The Public Client and the Construction industries (The Wood Report)*, London: National Economic Development Office.

31 National Economic Development Office (1970) *Large Industrial Sites. Report of a Working Party,* London: National Economic Development Office.
32 Building Economic Development Committee (1983) *Faster Building for Industry,* London: National Economic Development Office.
33 Cherns, A.B. and Bryant, D.T. (1984) 'Studying the client's role in construction management', *Construction Management and Economics* 2, 177–184.
34 Hewitt, R.A. (1985) 'The procurement of buildings: proposals to improve the performance of the industry', unpublished project report submitted to the College of Estate Management for the RICS Diploma in Project Management.
35 Bresen, M.J., Haslam, C.O., Beardsworth, A.D., Bryman, A.E. and Keil, E.T. (1988) 'Performance on site and the building client', occasional paper no. 42, Ascot: Chartered Institute of Building.
36 Building Economic Development Committee (1988) *Faster Building for Commerce*, London: National Economic Development Office.
37 Lindblom, C.E. (1971) 'Defining the policy problem', in *Decisions, Organisations and Society*, Castle, F.G., Murray, D.J. and Potter, D.C. (eds), London: Penguin Books.
38 MacKinder, M. and Marvin, H. (1982) *Design Decision-making in Architectural Practice,* York: Institute of Advanced Architectural Studies, York University.
39 National Economic Development Office (1978) *Construction for Industrial Recovery*, London: HMSO.
40 Nutt, P.C. (1984) 'Types of organisational decision processes', Cornell University, *Administrative Science Quarterly* 29, 414.
41 Mintzberg, H., Raisinghani, D. and Théorêt, A. (1976) 'The structure of unstructured decision processes', *Administrative Science Quarterly* 1, 246.
42 Bowley, M. (1966) *The British Building Industry. Four Studies in Response and Resistance to Change,* Cambridge: Cambridge University Press.
43 Pettigrew, A.M. (1973) *The Politics of Organisational Decision-making,* London: Tavistock Publications.
44 Bass, B.M. (1983) *Organisational Decision-making*, New York: Irwin.
45 Rowlinson, S.M. and Newcombe, R. (1986) 'The influence of procurement form on project performance', paper presented to the Chartered Institute of Building 1986 Conference, Montreal, Canada.

10 Successful building procurement system selection

10.1 Introduction

This chapter attempts to bring together all of the lessons learned from the previous examination of the existing literature and the past and current research relating to the successful selection of a building procurement system.

The success of a project can only be determined by the degree to which it meets the client's and/or the end-users needs and requirements. With a very competent and experienced project team that has worked together on many similar projects, the choice of procurement system will not always be critical to achieving client satisfaction. However, the majority of project teams do not exhibit all of these characteristics. Therefore, it is suggested that, in most cases, it is necessary to select the procurement system with extreme care by following the procedure described in previous chapters by:

- Identifying the characteristics and culture of the client and his/her organisation, establishing his/her needs and requirements, and those of the end-user of the facility, together with any internal and external constraints and risks associated with the proposed development and, of course, the project objectives.
- Ensuring that all of the currently available procurement systems together with their respective characteristics, advantages, disadvantages and means of implementation are identified and known to the person or persons who will be responsible for the selection of the procurement system.
- Choosing the appropriate system by matching the client and project profiles to the method of procurement with the most suitable and compatible characteristics.
- Monitoring the effectiveness of the selected procurement method by obtaining feedback from all members of the project team.

All of these elements of the procurement selection process are now considered.

10.2 Clients' characteristics, needs and project objectives

Clients' characteristics and culture

The culture of any client establishment is determined by its history and ownership, size, the technology it uses, its goals, objectives, environment and people [1]. For example, the informal, relaxed, casual dress management style of the typical, young, founder-dominated, US-based software house and the more formal, conservative ethos of a long-established, major international bank is likely to result in very different approaches to the way in which they each handle any major project despite the fact that they can both be categorised as private/experienced/secondary clients.

It therefore follows that although it is well established that the characteristics of the client and/or his/her organisation will affect the way in which project-implementation categorisation is approached, i.e. into which of the six previously defined categories the client falls, it can only be used as an initial guide to the likely characteristics of any client.

Any externally, or even internally, appointed project manager needs to be conscious that organisations are complex bodies and that, after placing the establishment into a particular category, it is also necessary to determine the specific characteristics, culture, policies and philosophies in order to refine the client's profile.

The complexity of many organisations, which incidentally is not necessarily always related to size, also makes it essential that the characteristics of the major departments, and their management, are identified, particularly if they are to be occupying or having an involvement in the proposed project. In the same way, details of outside interests – such as consortia partners, financial funders, major shareholders and insurers, all of whom may well have a direct or indirect influence over the organisation's approach to project implementation – should be identified and included in the profile.

It is easy to lose sight of the fact that in the case of many construction projects the client/owner usually only sees the end product and often has very little involvement with, or knowledge of, the actual project-implementation process. This is particularly true when the client is inexperienced, or only partially experienced, in terms of the implementation of construction projects. The principal adviser should then ensure that the client is made aware of the principles of the project-implementation process, the difficulties and dangers inherent in it and the various way in which the procurement of the project can be achieved.

The client's reaction to this information should provide an interesting insight into the culture of the organisation and the characteristics of its management, and should, hopefully, ensure that the client is less likely to be surprised when faced with the reality of the unique way in which construction projects are implemented when compared with the methods used to produce and provide products and services in the industrial and commercial sectors of industry.

The formulation of the client's profile, perhaps using as a base Handy's [1] cultural characteristics, is the first step to ensuring that any subsequent decisions are made knowing that they will be compatible with the organisation's policies, philosophies and management style.

Clients' needs and requirements

The client's requirements need to be established in advance of the project objectives and will not only embrace the spatial, functional and quality needs but also, most importantly from the point of view of the procurement system selection, the way he/she wishes the project to be structured, particularly in terms of risk, any internally or externally imposed constraints and any requirements stemming from the influence of outside interests.

Reference has already been made to the need to identify the real client when attempting to establish his/her needs and requirements and the difficulties that this can cause when the project is to be used to satisfy the needs of disparate entities within an organisation. On a project where the end-user is different from the owner, for example where the latter is a developer and the former a tenant, this situation is exacerbated and will make the determination of the needs of the eventual occupier a difficult, and sometimes near impossible, task.

Despite these difficulties, every effort needs to be made to establish what the eventual user of the project requires, although on occasion this can conflict with the objectives of the owner and diplomacy and tact will be required in order to establish priorities and/or a compromise between the differing requirements.

It is essential to ensure that the interests of the various internal factions are accommodated and that the final project brief reflects all of their requirements, or any compromises reached, and is 'signed off' by all of the interested parties.

Some clients' requirements will in themselves be constraints upon the way the project and especially the selection of the most appropriate procurement system can be carried out. As we have seen, the internal environment of any category of client organisation can impose constraints upon the project; this is most likely to be in the form of restrictions stemming from the policies or standing orders/internal regulations, or even unwritten rules, of the organisation.

While the most obvious examples of restrictions, such as the use of certain specific tendering strategies, types and forms of contract, have already been discussed, other restraints can range from restrictions on the use of certain consultants, contractors and procurement systems to the imposition of unrealistically low financial approval by authorities on clients' project managers.

All such constraints are likely to be to some degree detrimental to the successful and efficient implementation of the project, but many organisations, even those that appear to be extremely commercial in all their other dealings, are often not willing to waive any internal regulations. Some clients, possibly those in the private rather than the public sector, may be persuaded by an exercise which compares the cost of adhering to the policies, etc. with the lower cost that would be incurred if the restrictions were removed or amended. A similar exercise with time as the criterion might also prove fruitful.

In many cases, reluctance to amend or waive internal policies or regulations stems from the organisation being unprepared to accept any real, or perceived, risk that any divergence would create or worsen. The use of risk analysis techniques can be used to assess accurately the level of risk, thereby enabling financial/time comparisons to be made which might serve to persuade the client to adopt a less conservative approach to amending or waiving the restraint.

Project funding can often prove to be one of the more fertile areas for the production of constraints, particularly if the client is obtaining finance from an external source of funding that has its own particular expenditure programme requirements or that requires specific spatial, functional or quality standards to ensure market acceptability should disaster strike and the project revert to the funder's ownership.

Clients who are part of central, or in some cases local, government are well known for being restricted by various policies, regulations, etc., which are there to ensure public accountability, but they are just as often constrained by funding being restricted to a specific time frame, their inability to roll over funding from one year to the next, the drip feeding of funding for long-term projects and the general inflexibility of the central and local government funding systems.

As most funding-related constraints cannot be removed without great difficulty, it is usually only possible to mitigate the consequences by selecting methods of procurement which do not inhibit expenditure from being incurred at times when money is available or even for other members of the project team to provide short- or long-term finance, as with private finance initiative (PFI) schemes. However, such devices are likely to prove costly and may even detract from, reduce or possibly invalidate the viability of the project; as a result, these devices must be determined as part of the project strategy and incorporated into any requests for tenders or proposals.

Some physical constraints can affect the choice of procurement system, for example the presence of a number of existing occupied buildings on the site which are only going to be abandoned in stages and thus become available for demolition or refurbishment, could well determine the way in which the project is procured.

The question of risk in construction projects also needs to be considered at a very early stage in the project's life in order that risks can be identified and responsibilities for accepting such risks determined. These decisions will in turn determine which procurement systems are compatible with the level of risk that the client wishes to take upon himself/herself or pass on to others.

Flanagan [2] has identified the apportionment of financial risk between client and contractor as being dependant upon which procurement approach has been adopted for the project. This ranges from virtually 100 per cent acceptance of risk by the contractor when using the package deal system to the same rate of acceptance by the client when using the pure form of construction management. It can thus be seen that clients who are risk averse are considerably reducing the number of procurement systems that are available to them, whereas organisations that are prepared to accept risks that they are able to manage effectively themselves will have a wider range of systems to choose from.

Project objectives

In the heat of the project-implementation process, it is often forgotten that the object of the exercise is to satisfy the client by meeting his/her project objectives. This can best be achieved, as in most other human activities, by spending sufficient time planning and preparing the ground for the implementation process itself; the formulation of a considered and comprehensive project strategy will ensure that this is done and that success is finally achieved.

Historical evidence and much research demonstrates that clients in general are not particularly good at accurately and expeditiously determining their project objectives, with less experienced clients being worst at carrying out this task.

The project objectives will not only be determined by the specific requirements and aims of the project itself but will also obviously reflect the organisation's wider strategy and long-term business/social/political aims, i.e. the client's requirements.

Difficulties can arise during the project-definition stage when narrow project objectives conflict with long-term corporate goals. Departmental, divisional or subsidiary companies within large organisations or groups often have their own agendas and accidentally or deliberately overlook, or are not aware of, the wider corporate/political or social issues.

Conflict among departments, divisions or individual companies within a group can also create problems, and once again the need to identify the real client is of major importance when establishing the project objectives and satisfying the demands of disparate entities within an organisation.

The control of any activity needs to be carried out against a set of aims, or goals; in the case of construction activities, the project objectives. These objectives are usually communicated to the project team by means of a project brief. This document, as well as containing the client's functional, spatial and technical needs, should include details of the required timing, cost parameters and quality/functionality standards.

An accurate, definitive and timely project brief will enable an early decision to be made with regard to the choice of the most appropriate procurement system. Too often, an early decision is made on building projects to appoint a team of design consultants rather than a single unbiased principal adviser, with the result that a procurement system is often chosen by default rather than design.

The three primary objectives of time, cost and quality/functionality need to be closely defined in some detail so as to ensure that the specific project objectives are accurately identified.

Some clients, and principal advisers, may be concerned at divulging details of cost estimates, but little is lost and a great deal gained from all members of the team being aware of the financial parameters that have been set and the majority of competent and professional team members will objectively endeavour to better rather than just meet these targets.

To be told that the client requires minimum cost, which itself raises a number of issues, is insufficient. An actual maximum cost budget needs to be set, a time schedule for expenditure specified, a contingency sum agreed, accountability requirements defined, etc. as all of these requirements could impinge on all, or some, of the other elements within the project strategy. In the same way, expressing time in simplistic terms, i.e. 'the shortest time possible', is unacceptable. Project duration, the design period, the required commencement date, sectional handover, the length of time allowed for construction, the length of the defects liability period, etc. all need to be specified when they are appropriate to the client's project objectives.

Quality also needs to be expressed in specific terms using appropriate accepted industry standards in written specifications, sample materials and panels, room mock-ups, etc. The achievement of functionality will of course need to rely eventually on drawings and other visual aids, but, in the early stages of a project, performance specifications will be used to establish the client's requirements.

All of the three primary objectives need to be prioritised and weighted. This is a task which many clients, particularly those with little experience of project implementation, find difficult but which is vital to the eventual success of any project.

Once the primary objectives have been prioritised and weighted, a number of procurement systems will have been found to be inappropriate for use on the project under consideration; this number may be further reduced when the project's secondary objectives have been considered.

Although they are by their very nature individually of lesser importance than the three main objectives, in combination the secondary objectives can have considerable influence on the project procurement process. Many clients have, for example, a preference for carrying out the installation of their process or manufacturing equipment during the currency of the main construction contract. Whereas in practice this is normally perfectly possible, problems can arise if the client's requirements in this respect have not been fully considered during the selection of the chosen procurement method and then accurately and comprehensively incorporated within the tender documentation and subsequent contractual arrangements.

Should the client wish to be able to operate his/her facility with a trained staff immediately it has been handed over by the contractor, wish to deal with only one single organisation during the project duration, wish to use specific design professionals during the early stages of the project and then place the responsibility for detailed design and construction onto a contractor, etc., these apparently secondary requirements will have a significant effect on the category of procurement system that he/she will be able to use.

Should two or more of these examples of secondary objectives be required in combination on a single project, they would be likely to have such a major effect on the procurement process that they should be redesignated as primary rather than secondary. The client's secondary needs should also be prioritised and weighted.

The proper and early identification of the client's objectives is a vital step in formulating the project strategy and selecting the most appropriate procurement system, thus ensuring the effective control of the procurement process and the management of the project as a whole. Time spent on ensuring that objectives are properly and accurately identified at this stage will be repaid many times over during the implementation of the project.

10.3 Identifying all currently available procurement systems

Research by the author has shown that very few clients, even those that have had experience of building, are aware of all the procurement systems or the way the systems themselves operate. There is also some evidence, admittedly partially anecdotal, which suggests that the average consultant is also unaware of the full range of potential methods or their characteristics.

We also now know that many clients appoint their design consultants or contractors during the initial stages of the project, when such pre-emptive appointments can only hinder, or make unworkable, the unbiased selection of the most appropriate procurement system; the appointment of such organisations should only take place once the procurement route has been chosen. It is therefore suggested that clients should look to the, hopefully unbiased and independent, person whom they have appointed to advise on, or carry out, the initial management or co-ordination of the project to identify all of the suitable procurement systems.

This exercise will entail not only identifying all of the appropriate methods but also obtaining details of the way the individual methods are implemented, the characteristics of the various systems and their advantages and disadvantages. This information is now quite readily available in the literature, from various sources on the internet, from government departments, from many of the construction industry's representative bodies and from professional institutions.

Ideally, a short report, or discussion paper, containing all of this information should be prepared for, submitted to and discussed with the client in order to make him/her aware of the methods that will be considered at the next stage in the process and to obtain his/her reaction to the characteristics, etc. of the various systems.

Once the final list of potential procurement systems has been agreed, it will be time to move on to the most critical stage of all – selection of the most appropriate method.

10.4 Choosing the most appropriate procurement system

Research has shown that many clients, possibly the majority, do not select the procurement method they eventually use in a disciplined and logical way.

If success is to be achieved with minimum difficulty, the decision regarding the most effective means of procuring the project must be made by comparing the client's characteristics, needs and project objectives with the characteristics, advantages and disadvantages of each of the selected systems; whichever of the systems is most compatible is then chosen.

Despite the documented reluctance of clients to use the aids to selection that have previously been described, these techniques can be of great assistance in reducing the number of potential procurement systems to more manageable numbers and, in some cases, to select specific methods as being most suitable for the project under examination.

It is suggested that the manual method of selection designed by Skitmore and Marsden or the computerised ELSIE system are most likely at the present time to produce the best results. Love *et al.* [3] found that such methods of incorporating a simple set of criteria are generally adequate and sufficient to enable a successful procurement path selection to be made.

Once again, it must be stressed that these and any similar aids should only be used to provide guidance or confirm preliminary decisions and that the final choice will always need to be made by the responsible party, who will need to take into consideration external issues such as the state of the economy, availability of desired project team members, tenderers, etc.

Once the choice has been made, the client will need to be advised not only of the decision but most importantly of the contractual, practical and financial consequences of using the selected method, in order that formal approval may be obtained before any further action is taken.

The forecasting of the likely decrease, or increase, in the cost to the client of using different procurement systems has always proved to be difficult. The 1996 report *Designing and Building a World Class Industry* [4] demonstrated that projects carried out using the design and build system were at least 13 per cent cheaper than similar projects implemented by conventional methods. Apart from this work, little progress was made in this area until the University of Manchester Institute of Science and Technology's (UMIST) research project [5].

The UMIST research produced a neural network cost model, which is still under development, that as well as predicting the total cost to the client per square metre of a project will enable a comparison to be made between the costs of using different procurement systems.

Once the final selection of the most appropriate procurement system has been made, the client needs to consider the nature of the environment in which he/she wishes to implement the chosen method. Does he/she wish the project to be carried out within a climate of co-operation or confrontation?

Even under the comparatively ideal circumstances encountered in manufacturing industry and commerce, persuading individuals or disparate groups to work together successfully towards a common goal is never an easy task and demands the presence of supportive policies, above-average management skills and an organisational structure that is both flexible and capable of dealing with cross-departmental operations and the requirements of matrix management.

In the not too distant past, construction projects have been totally unable to meet these criteria because of the fragmented nature of the project organisation, the lack of decisive and strong leadership, the rigid nature of most contractual arrangements and the confrontational nature of the industry as a whole.

The fact that project teams in construction usually consist of individuals from different organisations, each with its own agenda and objectives, who are more often than not unknown to each other would be sufficient on its own to account for the difficulties experienced in managing projects. But in addition to this constraint, the tendering arrangements, conventional procurement systems, disputatious types and forms of contract, the unique nature of construction projects, the difficult physical environment in which work is carried out and the combative attitude of many managers mean that the establishment of a positive, co-operative and supportive approach by members of the team is extremely difficult.

Clients have a very important part to play in overcoming these difficulties, initially by establishing an environment in which co-operation can flourish and in which positive, objective attitudes are encouraged. Even without the use of non-confrontational and/or unconventional contractual arrangements, a positive environment can be realised by any client organisation willing to adopt appropriate policies and attitudes and to ensure they are communicated, established and adopted throughout the project team, and in every aspect of the project's life, in order to create a constructive project culture.

This has always been achieved on a relatively small number of projects by the simple expedient of the client being constructively and continuously involved with the implementation process and by applying the same enlightened philosophies and approaches as those adopted within his/her own organisation coupled with the use of tried and trusted consultants, contractors, suppliers and subcontractors and utilising the most appropriate procurement systems and contractual arrangements.

Where such an approach has been used, the project has been successful because the client has always been satisfied that his/her objectives have been met. Also, all the other members of the team have benefited by achieving their short-term commercial goals and the long-term aim of satisfying an important client and heightening the chances of their services being used on future projects.

Unfortunately, such far-sighted and sophisticated clients, or project teams, are not common in the construction industry. In order to achieve the same level of success more generally, it is necessary to formalise this enlightened approach into some form of arrangement similar to that now known as 'partnering' and/or the use of non-confrontational procurement systems such as those within the management-orientated category.

These techniques have been shown to produce a 'can do' attitude among the participants and to foster and require a co-operative approach to problem-solving among the client and the team members, and, in conjunction with the use of the correctly chosen procurement method, help to ensure a successful project.

Communicating the choice of procurement system once it has been made to all of the parties that are participating in the project at this stage is often overlooked, but the inclusion of the decision within the circulated project brief should ensure that all of the participants are aware of the choice that has been made.

Depending on the nature of the client organisation, and such external members of the project team that have been appointed, it is often of benefit that the consequences of the choice in terms of the advantages and disadvantages of its use and the constraints that it places on the behaviour and involvement of the current participants in the future implementation of the project should be spelt out to all concerned and confirmed in writing if felt appropriate.

10.5 Monitoring and feedback

All commercial activities need to be monitored in order that progress can be measured against specific objectives, and if necessary remedial action can be taken to correct under- or overachievement.

The monitoring of the effectiveness of the chosen procurement system presents some difficulty because of the qualitative nature of the task. The majority of the primary and secondary project objectives are capable of being measured against quantitative targets, but whether or not the choice of procurement system has been correct cannot be determined other than by the measures of success applied to these objectives and the final outcome of the project-implementation process.

During the implementation of the project, it is of course possible to obtain feedback on the way the system is performing from individual members of the project team, although it will be appreciated that such information may be subject to personal and professional bias. Notwithstanding this caveat, regular monitoring of the team members needs to be carried out against a formal set of criteria and their comments noted.

The client's impression of the way in which the system is performing is of primary interest and importance and must therefore be monitored and recorded in the same way as the opinions of the project team. At the completion of the project, the initial criteria for success reflected in the client's needs and the project objectives should be the benchmark against which the performance of the project team and the procurement system itself should be measured.

The feedback obtained from the ongoing exercises will provide valuable information on the way in which the system is performing and will enable the project team to establish whether its performance is satisfactory; unfortunately, should this not be the case it is rarely possible for the choice of procurement method to be changed during the currency of the construction phase of a project without experiencing considerable difficulty in making the necessary contractual and practical modifications. Under such circumstances, any necessary changes will need to be taken by addressing the design, construction methods or management elements of the project.

Notwithstanding such difficulties, the guidance obtained from feedback as to the management of the procurement, and other aspects of the project, together with all of the information gathered over its duration will be extremely valuable and, once completion has been achieved, should be incorporated into a brief report and circulated to all members of the project team.

References

1   Handy, C.B. (1985) *Understanding Organisations*, Harmondsworth: Penguin.
2   Flanagan, R. (1981) 'Change the system', *Building* 20 March.
3   Love, P.E.D., Skitmore, M. and Earl, G. (1998) 'Selecting a suitable procurement method for a building project', *Construction Management and Economics* 16.
4   Bennett, J., Pothecary, E. and Robinson, G. (1996) *Designing and Building a World Class Industry*, Reading: Centre for Strategic Studies in Construction.
5   Harding, A., Lowe, D., Hickson, A., Emsley, M. and Duff, A.R. (2000) 'Implementation of a neural network model for the cost of different procurement approaches', in *Proceedings of the CIB W92 Procurement System Symposium – Information and Communication in Construction Procurement*, Santiago, Chile, 24–27 April, Santiago: Commission for International Building.

11 Future trends in project procurement

11.1 Procurement in general

Both the demand and the supply sides of the construction industry have the ability, and in the case of some organisations the desire, to shape the future.

However, in reality the large experienced corporate client is more equal than the other members of the project team in this respect, and has demonstrated this over the past three decades by introducing and using procurement processes of his/her choosing rather than accepting what the industry had to offer at the time.

The use of management-orientated procurement systems on major projects, integrated methods of procurement on many medium-sized and small developments, the short-lived British Property Federation (BPF) system, the public clients' increased use of private finance initiatives to fund, construct and operate projects and now the increasing use of partnering, the Ministry of Defence's experimental 'prime contracting' approach and Slough Estates' 'new design and build' system all have stemmed from client demands or initiatives.

Major clients are now however going further and making specific demands upon the industry. In March 1998, the Construction Clients' Forum, an informal grouping of fifteen of the largest construction clients accounting for about 80 per cent of the UK construction market, published *Constructing Improvement* [1], essentially a list of actions required from the supply side and promises made by the demand side which, the organisation believes, should lead to the improvement of construction project performance.

Focusing on the four elements of the procurement process that are being examined in this work, the client's requirements can be summarised as:

- focus on achieving results for the client and meeting client's objectives;
- invest in research to improve the chances of meeting client's needs;
- improve management of the supply chain;
- provide more value than clients expect;
- work towards more standardisation;
- provide clarity of advice about available options during the implementation process;

- provide comprehensive information about problems and the implication of changes.

Clients in return promise to:

- set more realistic targets;
- communicate objectives and decisions more clearly;
- promote relationships based on teamwork and trust;
- apportion risk sensibly in their project contracts, which will be standard wherever possible;
- improve their own management and construction techniques;
- promote co-operation.

So, what clues can be found in these specific proposals, the Egan Report [2] and other contemporary literature in general to the way in which the pre-tender stages of the project procurement process will develop in the future?

Project definition and the establishment of the client's objectives should become increasingly more disciplined as clients realise the commercial dangers of not planning their projects in a systematic manner during the initial stages; in other words, the formulation of a project strategy and a detailed plan and programme for the implementation of all the stages of their development should become far more common than at present.

If accomplished, the aim of upgrading the client's own management standards and expertise should enable this improvement to be achieved particularly when allied to the proposed clearer communication of objectives and decisions to the other members of the project team.

The need for better communication between the various parties in the whole of the construction process cannot be sufficiently stressed. Historical evidence from major military, commercial and industrial projects shows that the lack of comprehensive, accurate and speedy communication of the client's objectives and decisions, as well as any changes to them during the currency of the project, has been one of the main reasons for time and cost overruns.

The setting of the three primary objectives of time, cost and quality/functionality needs to be carried out as part of the formulation of the client's project brief. At the moment, this if often not done, but the promise by clients to set more realistic targets should also heighten future awareness of the need to include this vital information in this essential document.

One of the existing major constraints to ensuring a successful conclusion to a project is the attitude adopted by the supply-side members of the project team, who tend to be focused on its own objectives rather than those of the client and his/her project. If, as clients have demanded, this negative attitude can be changed and a 'can do' co-operative approach adopted instead, the chances of future success will undoubtedly be substantially enhanced.

The requirement that the supply side invests in more research than at present is apposite as the majority of current research work is being funded by central government and client organisations and carried out by academia.

To ensure that this request is implemented may well prove to be more difficult as consultants and contractors have a poor record for funding research projects, and it is therefore doubtful as to whether there will be a significant increase in the amount of research commissioned by such organisations in the future.

The client's undertaking to promote co-operation and trust among all members of the project team together with the improvements requested in the management of the supply chain reflects the current enthusiasm among employers for partnering-type arrangements. This is also reflected in the clients' need for clarity of advice about options and their request for information on problems and the consequences of changes to be made more readily available during the implementation process. Such needs can only be totally satisfied when there is an environment in which all the parties in the process have everything to gain, and nothing to lose, by expressing opinions and providing information.

In the immediate future, there is little doubt, particularly with major corporate clients, that the use of this more collective approach will continue and increase. In the longer term, and especially in times of reduced workloads, the co-operative philosophy could be undermined by the short-termism that has been applied in the past by both sides of the industry to other new and worthwhile techniques and methods that have been introduced to the construction sector in order to improve its performance.

For example, the demand by many clients for a guaranteed maximum price when using management-orientated procurement systems in which the manager is supposedly appointed as a construction consultant and the unfair treatment of trade and package contractors by some management contractors and construction managers could well be repeated when the economy is in recession and competition in the construction industry is intense.

The promises by the demand side to apportion risk sensibly and to use standard forms of contract should tend, if they actually come to fruition, to mitigate against the worst excesses of future commercial avarice in this respect, but it is likely that, as a result of economic pressures, the use of co-operative approaches may well be reduced or detrimental alterations made to the method for the benefit of clients or contractors.

Much has been made by client organisations recently of the need for greater standardisation in the industry, with comparisons with the automobile industry being constantly made. For different reasons, some successful attempts were made in the UK during the 1960s to standardise, or systemise as it was then known, the construction process. Although favourable – if comparatively short-lived – results were obtained at that time, they were mainly restricted to the mass housing market in both the private and public sectors.

Those commercial and industrial systems that were developed were small in number and tended to be highly technical, and expensive, in their application, and the majority did not survive the decade. The fact that packaged, predesigned standard buildings have not captured a larger share of the market reflects the dislike of most clients for standard designs because of the restrictions they impose on the aesthetic and functional aspects of a project.

While it would undoubtedly be advantageous if more standardisation was realised, it is suggested that it is more likely that an increase in the use of standard components and the off-site manufacture of larger elements of the project may well be the best that can be achieved.

The use of lean production methods in construction will increase as a result of pressure from government and major private clients following the report [2] from Sir John Egan's construction task force. Lean production techniques are now well established in most industries worldwide, although remaining at the trial/experimental stage in construction. The method is closely linked to quality management and requires all operations to be 'right first time', thereby eliminating reworking and any activity that does not add value to any stage or element of the implementation process.

The task force wants the construction industry, in addition to increasing the use of partnering, to reduce construction periods by half, to achieve a 50 per cent reduction in defects and to eliminate waste and waiting time.

Initial reaction from all sides of the industry has been somewhat scathing and incredulous at the size of cost and time reduction targets that have been set, and it is suggested that although the more proficient contractors and consultants are likely to go some way to meeting the targets the majority of the supply-side organisations will have difficulty in achieving the goals that have been set.

We have seen that clients in general, but particularly those with major construction programmes, have become increasingly sophisticated over the past quarter of a century. In addition to the demands that have already been discussed, there are signs that they are beginning to expect more accurate forecasts of the likely comparative project costs when using alternative procurement systems.

While the various aids to procurement system selection that have already been described enable the client to identify the most appropriate method of procurement based upon his/her needs in terms of speed, quality and functionality, very little help exists to enable any comparison to be made of the likely project cost when using the various systems.

It has already been noted that in 1996 The University of Reading in collaboration with the Construction Clients' Forum produced a report [3] in which the use of the design and build system was found to be cheaper by up to 30 per cent than conventional methods.

However, apart from this research, very little appears to have been done to determine such comparisons until a successful study initiated by the

Department of Building Engineering at the University of Manchester Institute of Science and Technology in 1997 established the feasibility of collecting the necessary historical cost data, analysing them and developing a comparative cost model.

Work is now proceeding on the second stage of the project, and within the foreseeable future a computerised predictive cost model of all the main procurement methods should be available.

11.2 Procurement systems in particular

With clients becoming increasingly involved in the overall management of their projects, the number of procurement systems is likely to grow as experienced clients continue to introduce variants of the existing methods and to experiment with entirely new ways of procuring their projects. On the other hand, less experienced clients are likely to continue to choose their procurement systems in a haphazard way from a narrow range of the numerous methods that will be available in the future.

The future level of use of the individual procurement systems that have been previously described can only be based upon current and past usage and a little educated conjecture. The most definitive data available are those produced for the Royal Institution of Chartered Surveyors (RICS) in its biennial *Contracts in Use* survey – referred to many times within this text. These data would appear to indicate the following future trends.

The *conventional* system has seen a substantial decline in use since the mid-1980s, when measured by project value, and a reduction, although less dramatic, in the number of projects carried out using this method. It is suggested that its use on major high-value projects will continue to reduce, although perhaps at a lower rate than in the past, whereas on medium and small projects there is the possibility that the decline will continue more slowly and might even stabilise.

Despite anecdotal evidence suggesting that the *design and build* system has been used inappropriately in the past, the use of this method has continued to increase very substantially over the past decade, both in terms of value and number. This obviously reflects the desire of clients to deal with a single entity when implementing their construction projects while at the same time enjoying the advantage of, in theory at least, being in possession of a guarantee to complete the project for a specific sum.

The perceived advantages of this system are such that its use will probably continue to grow, although it is difficult to envisage such a dramatic future increase as has occurred over the past few years. The fact that the method is no longer seen as only suitable for very simple utilitarian buildings is likely to ensure its use on an increasingly wide range of projects.

Although severe criticism of the *management contracting* system during the early 1990s appears to have resulted in a reduction in the use of this method from that time until the mid-1990s, the latest statistics indicate a return to favour during the last few years of the twentieth century.

The fact that the 1998 RICS survey found that only 50 per cent of the projects carried out by this method used the Joint Contracts Tribunal (JCT) management form of contract, with the remaining percentage preferring contracts written by clients, quantity surveyors or contractors, could suggest that clients were using the system to ensure that they were able to penalise the management contractor should the project fail to be completed on time or to budget. Contractor-designed contracts could be used to protect their interests in this respect and to ensure that they were not restricted in their ability to control payments to works contractors as they wished.

Should either of these assumptions prove to be correct, such regression to past short-term behaviour, while being detrimental to the proper use of the system, might well mean that its future use could well increase to the levels prevalent in the 1980s.

There was a modest increase in the use of *construction management* during the latter part of the 1990s, although the level of use is still less, whether measured by value or number of projects, than management contracting and its use still appears to be limited to major, high-value and/or complex projects.

It is suggested that as these criteria are likely to continue to be the main reasons for choosing this particular system there is unlikely to be any dramatic increase in the use of this method, although, if past trends are repeated in the future, there should be some moderate growth in the employment of the system. The high value of most construction-management projects can of course produce wide fluctuations in the year-on-year levels of use, and this characteristic will continue to be evident until such time as the system is more widely used on medium-sized, lower value projects.

The long-term future of the system can only be seen as being good, with the positive 'can do' attitude that should be engendered by use of the method pleasing clients and ensuring its future success.

The lack of up-to-date information on the past and current use of *design and manage* makes any forecast of the future of this system extremely difficult, and it can only be suggested that in the light of the ability of the method to deal with a wide range of projects it will continue to be used at about its present level.

Once again, there are no definitive data available on the use of the *British Property Federation* (BPF) system, even from the Federation itself, but all the indications and anecdotal evidence suggest that the use of this method, even among members of the organisation which initiated it, has been less than originally envisaged by its authors. There is, therefore, no reason to believe that the future will see an increase in its use and the fact that the BPF no longer publishes the contract documentation could suggest that its demise, at least in its original form, is only a matter of time.

Finally, the future of *partnering* – the survey carried out by the RICS of contracts in use for 1998 [4] – captured, for the first time, the number of contracts incorporating a partnering or alliance provision, although only forty-two of the 2,457 projects surveyed, i.e. 1.7 per cent of the total, had such a provision.

Despite this apparently low level of use, all of the non-statistical evidence suggests that the use of this 'administrative' method is common on many major projects, reflects the cultural changes in the industry that are being sought by government and large commercial clients and is proving to be beneficial to all of the members of the project team.

All of these factors seem to indicate that the level of use of this method will continue to grow as more clients, consultants and contractors experience its advantages, including the ability to be able to be used on single projects as well as long-term construction programmes.

11.3 Summary

It should be stressed that no attempt has been made in this chapter to do any more than suggest the possible scenarios for the future development of the procurement process in general and the individual procurement systems in particular.

There is little doubt that clients and perhaps the industry itself will continue to produce or demand new methods of procuring projects. These new systems, or variants of existing systems, will be designed to provide solutions to the problems or unique circumstances of specific development proposals. If they are successful in overcoming specific obstacles and are well publicised, they will become fashionable, be taken up by others and used on inappropriate, as well as appropriate, projects, and will eventually either become part of the portfolio of tried and trusted systems or disappear into obscurity.

In general, it is suggested that further demands on the supply side to reduce design and construction times and costs will be made by a more disciplined body of clients, who will be defining their project objectives more accurately and communicating them more effectively than at present.

There will hopefully be an increase in client involvement in all aspects of the project, resulting in the establishment of a project environment which helps to establish trust among those that operate within it and engenders more co-operation among team members and encourages the reduction of confrontational attitudes.

Some increase in the use of standardisation is envisaged but there is likely to be much more emphasis on lean construction, although the targets for improvement that have been suggested appear to be overly optimistic.

Specifically, there will be better definition of the client's objectives, particularly on major projects; a reduction of non-physical client-generated constraints; the better management and more equitable distribution of risk; and, as a result of improved management and the availability of more sophisticated selection techniques, a more disciplined and systematic choice of procurement system.

Having said all of this, it is suggested that success is more dependent on establishing an appropriate project environment in which the project team can deliver the client's project objectives than any particular technique or system of procurement.

Improvements will also stem from the elimination of the automatic acceptance of the lowest tender, which remains the policy of many client organisations, and instead the appointment of contractors being based upon the bid that provides the best value for money for the client.

In the future, as today, only people working together in harmony with mutual goals and a sense of common purpose will produce successful projects and satisfied clients.

References

1 The Construction Clients' Forum (1998) *Constructing Improvement*, London: Construction Clients' Forum.
2 Department of Environment, Transport and Regions (1998) *Rethinking Construction, The Report of the Construction Task Force (The Egan Report)*, London: HMSO.
3 University of Reading (1996) *Designing and Building a World-class Industry*, Reading: Centre for Strategic Studies in Construction.
4 Davis, Langdon and Everest (2000) *Contracts in Use: A Survey of Contracts in Use During 1998*, London: Royal Institution of Chartered Surveyors.

Index